The Developmental Biology of *Urechis*

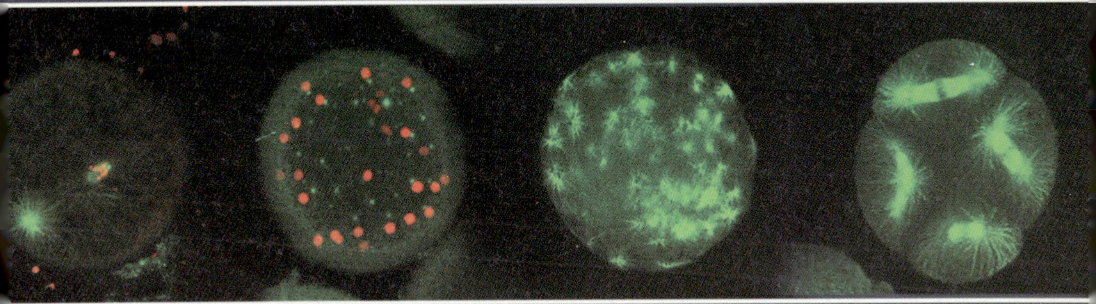

Kil Sang Shin, Ph. D.

SOON CHUN HYANG UNIVERSITY
PUBLICATION DEPARTMENT
SCH

The Developmental Biology of *Urechis*

First published 2011 by Poong Sam Sohn, the President of Soonchunhyang University

Soonchunhyang University Publication Department
646 Shinchang-myeon, Asan-si, Chungnam-do, 336-745, South Korea

336-745 KOREA
Tel: 82-41-530-1115
Fax: 82-41-530-1719
www.sch.ac.kr

ISBN: 978-89-966347-8-2 93470
Price: ₩ 30,000

This book dedicated to all of the many friends and colleagues whose help and advice over the years has enabled me to extend the scope of this book beyond my own limited experience.

PREFACE

At the beginning of the modern times when science was just another branch of philosophy, some were trying to find the Absolute using the essential elements of the quickening science. And thus science and philosophy separated their own ways, which caused some scientists to doubt the Absolute for Its existence and science became a mere tool to conquer the nature. The unfortunate belief that the nature was the infinite resources became widespread and people were urged to exploit and pollute it with newfound scientific knowledges. But it was discovered that the nature was being destroyed and they began to realize that the nature was something needed to be taken care of.

My students and I thought we, as scientists, contribute to the conservation of endangered innkeeper worms whose habitat was being threatened by rapid development. As times went on, I began to realize that innkeepers have been living, surviving, and helping themselves without any scientific knowledges and our helps. I also come to question whether motivations of many scientific endeavors were based on the conservation of the nature or economic gains. It takes quite a long time to grasp the deeper meaning of an old saying, “The boys get old easily and study is hard to achieve.”(少年以老 學難成). Science was developed over several millennia and only few finding researches still remain valid such as the principle of leverage from the prescience era and the speed of light. The fact that only few finding researches remain as “truth” may tell us that the scientific "truth" is nothing but what people believed to be "true" and its "truth" is created by how to make people to believe the natural phenomena. What we found as truth changes with times.

If the present research methods are also based on finding “why” and “how” in the nature through “touching” and “looking,” the results are bound to have errors and mistakes. They can be manipulated as well and the conflicting theories can be patched up together if scientists are more interested in something other than the quest for the truth. In this era where science is inseparable from commerce and politics, we just merely hope the

PREFACE

efforts of scientist are not wasted. However, I find solace and hope in my students. My students prove the theories contained in this monograph as facts by extensive experiments throughout the years. Innkeeper worms are the great materials for students to see the central theme of the life science. Innkeeper worms provide them with the current scientific languages. They are also great sources for hopes and passions. All of my students move on and become well adapted in their own field of studies after the life of university. I regret that all their works cannot not be included in this monograph.

We are in the middle of the third Scientific Revolution. With the new technologies and methodologies of the future, I hope there will be scientists who find this research of innkeeper worms interesting and worthy to pursue further. I hope this book to be a new starting point of their researches and those who read it will condone its imperfections as well as other errors which may hinder the way of efficient and rapid exchange of scientific information. Finally, I would like to thank my daughter On-sub, her husband HJ Jeon, and my wife SY Yi for editing, proofreading, and assisting many other tasks; without their help and support, this book would never have been completed. It is of course a pleasure to give special thanks to my colleagues, Dr. WJ Kim and Dr. MD Han for their generous help and support in bringing this work to fruition.

August, 2011. **Kil Sang Shin, Ph. D.**

CONTENTS

CONTENTS

CONTENTS

CONTENTS

CONTENTS

Chapter 01

The Biology of *Urechis unicinctus*

Chapter 01 The Biology of *Urechis unicinctus*

1-1 Classification

Urechis, marine metazoan with flexible and turnable bodies and proboscises, commonly called spoon worms (innkeeper or plump worms), live in the soil of tidelands. They are traditionally classified as Echiuroidea due to their unsegmented, coelomate bodies along with their unique anal sacs, proboscises, and muscle layers (Newby, 1940). *Urechis* are divided into 11 genus and 19 species (Sato, 1939). Three species, *U. unicinctus* (von Drasche), *U. chinensis* (M. Mueller), and *U. caupo* (Fischer), is more numerous compared to other species. *U. unicinctus* are distinguished by two pairs of gonoducts (sex organs) instead of three pairs typically found in other species, and is widely distributed along the coasts in Korea, Japan, and China (Sato, 1939).

Some authors have classified these worms as echinoderm while others place them into the phylum Annelida. In fact, these worms possess different characteristics of various marine animals, echinoderms, annelids and mollusks, *Spiralian,* and flatworms. Their external features and habitation resemble those of annelids, while their internal physiology including sexual characteristics is similar to that of mollusks and *Spiralians. Urechis* embryonic developmental patterns are similar to those of echinoderms.

Recent molecular data comparing the sequences of elongation factor-1 alpha genes support the hypothesis that members of this taxon are annelids (McHugh, 1997). Annelida, which includes polychaetes and clitellates, has long held the taxonomic rank of phylum. Unsegmented, mud-dwelling echiurans and gutless, deep-sea pogonophoran tube worms share several embryological and morphological features with annelids, but each group is also considered to represent separate metazoan phylum based on the unique characteristics of each group (Rouse & Fauchald, 1997; Cho *et al.*, 2006).

1-2 External morphology

U. unicinctus is spindle, or streamlined, in shape and unsegmented although segment rudiments can be observed. Middle aged worms are approximately 12~15 cm in length and 3 cm in width (Fig. 1-1). These organisms have smooth, slippery skin covered with abundant mucous substances that also coats the entire surface of burrow-walls for protection against the

Fig. 1-1 *U. unicinctus* on habitable tideland of the Mongsan-po beach. The adult worm is measured 12~15 cm in length and 3 cm in width. The average life span is expected 3~4 years. In the figure, the proboscis is split at the lower part of the left worm (arrow) and anus is round with white band of the lower part of the right worm (arrow head). One or two crabs (block arrow) are always found in burrow of the worm.

corrosive flow of sea water.

The colors of the worms vary from pale to dark brown, but tend to change according to the color of the soil in their habitat. The coloration of worms collected from the mud-sand of Mongsan-po is dark to blackish brown. Worms found in other part of Korea, such as the southern coast, are often bright brown in color. The front of the worm is more slender with a split proboscis while the hind portion is blunt with a pale, ring-shaped band at the tip of the tail. The split oral proboscis is short with a scoop. A pair of bristles or saetae protrudes ventrally from the proboscis, and bristles form a ring at the tail tip.

The muscular movements of the worm are exquisite. Flexible and resilient movement is possible as well as peristaltic contortions with repeated and continuous cycles of contraction and relaxation. The skin is glabrous and often makes it difficult to handle the worm. A method of anesthesia or narcotheraphy is not yet available currently for facilitating working with the worm. *U. unicinctus* is a coelomate with distinct genders. The sexes are not easily distinguished out of the breeding season. During the breeding season, the body (coelomic) cavity is fully filled with well-developed storage sacs: two pairs of yellow sacs in females and white sacs in males. At this time, the body wall becomes so thin that the massive outlines and coloration of the storage sacs can be observed, and can be used as a reference for distinguishing the sexes. After the breeding season, the storage sacs disappear completely and it is difficult to determine the gender even when the worm is opened.

Sexual activities of *U. unicinctus* are distinctly regulated by seasonal cycles. Sexual activity begins at the end of January and peaks at the beginning of May, during which time spawning and mating occur. The higher breeding season varies depending on the temperature of the seawater. During the summer, the worms are hidden 80~90 cm below the surface and remain there without any activity until mid-September. Some activity can be observed after this time, and vigorous activities are observed from December to January of next year. It is regarded that this worm goes into aestivation during the summer.

1-3 Habitat and the burrow

1-3-1. Habitat

The worms used for this study were collected from the west coast of South Korea, Mongsan-po beach/Anmyeon-do (an island)/Chungnam between 36° 34' 15.74" and 36° 32' 24.07" (N.L.), and 126° 18' 51.77" and 126° 19' 43.16" (E.L.). The total length of the beach was about 5 km with a width over 800 m at low tide (Fig. 1-2). From the hillock 10-25 m high that surrounded the beach, a large amount of fresh water flowed downward towards the beach, especially during the higher breeding season (May), mixed with seawater to form pools of brackish water with 16-22% salinity.

Ecological conditions of the tideland seem extreme due to changeable weather conditions, temperatures, and salinities of the brackish water. For example, salinities varied and ranged from15 to 22‰, during the breeding

Fig. 1-2 Monsgan-po beach: Right side is hillock of the beach and left is surface of the ocean. The breeding season is corresponded with the mild rainy season and an amount of brackish water flows over the tideland surface. The Mongsan-po beach is about 800 m wide at a lower tide and 5 km in length.

season to over 35‰ in midsummer. Seawater temperature ranged from 6-25°C, and great and heavy billows frequently broke down the soil of the beach. However, since the worm inhabits underground burrows, they can apparently avoid some of these unfavorable environmental conditions.

The coast provides the necessary and sufficient conditions that promote worm survival. These advantages may include the brackish water with lower salinity, depositions of coastal soil composed of fine sand and mud with an approximate ratio of 1:1, plentiful food including debris of various tideland animals, marine microorganisms, planktons, and young larva of various tidal animals hatched in accord with the worm's season, and the worm utilized apparently the larva as food materials.

Brackish water with lower salinity is especially important for improving the chance of oocyte fertilization (*See* fertilization). The oocyte is able to change into its proper shape with appropriate salinities. In brackish water with lower salinity (16-22‰), oocytes become round and expose a larger number of sperm receptors, microvillous tips, over the vitelline layer. On the contrary, exposure to brackish water with higher salinity (>22‰) causes indentation of the oocyte cytoplasm that conceals sperm receptors. When the salinity further increases (> 35‰), the oocyte deflates which is very difficult for the oocyte to become rounded.

1-3-2. The burrow

U. unicinctus is a type of cave dweller or burrower which bores into siphon-like burrows by itself under the tideland. Rather than sandy soil, this worm prefers to live in mixed soil composed of mud and sand, in which the burrow is constructed and the worm can find sufficient amounts of food. The soil of Mongsan-po tideland is blackish yellow in color and contains a mixture of yellow sand and blackish mud of approximately equal quantities. A region of pure sandy soil is found in a part of Mongsan-po beach in which worms are infrequently found.

The worm crawls about the burrow and secretes mucous substances onto the wall of the burrow to make it glabrous. The mucous secretions bind the mud-sand mixture together, turning it into sticky soil to maintain a proper burrow frame. Once the burrow is constructed, the burrow wall is resistant to erosion by seawater. It is difficult for the mucous secretions to turn pulverized sand lacking mud components into sticky soil. Worm burrows often collapse because the stickiness of the soilis not sufficient to resist damage from tides and billows. Worms living in mud-sand burrows can be captured with some effort, but worms inhabiting pulverized sand soil cannot be easily caught. This is because a mixture of sand and seawater easily flows inward into the excavated burrow (Fig. 1-5).

The worm turns freely to the front and rear of the burrow; it crawls about the burrow but usually orients its tail towards the efflux pore (Fig. 1-3) and proboscis towards the afflux pore (Fig. 1-4). During the breeding season, the proboscis is elongated with a smaller diameter and is deeply inserted into the narrow afflux pore until it reaches the tideland surface beneath. The inflow of sea water contains food materials that are filtered in the foregut or esophagus of the worm. The remaining seawater gently flows out through the afflux pore.

Food filtering is accomplished with cyclic inflow and outflow of sea water through the foregut or esophagus and afflux pore of the burrow. Rapid and acute sea water inflow is followed by gentle and slow outflow. As mentioned above, the cyclic movements of sea water are responsible for the characteristic shapes of the burrow structures. Respiration is accompanied by the cyclic inflow and outflow of sea water through the hind gut and efflux pore. Sea water in the burrow reservoir flows into the hindgut of the worm with negative pressure created by muscular expansion. The sea water is then discharged from the hind gut and through the efflux pore after gas exchange. Hence, the outpouring of sea water from the reservoir is always unidirectional, coupled with replenishing and refreshing of the reservoir water. Outpouring water is frequently spouted out, creating water columns over the tideland surface.

1-4 Body cavity and internal organs

Urechis is a coelomate with a well-developed body cavity created between body wall and internal visceral organs. The contents of the body cavity alternate with seasonal changes. Storage sacs and nephridia of both sexes appear and disappear according to different seasons.

When the worm is opened lengthwise out of breeding season, coelomic (body) fluid, coelomocytes, single lineup of the digestive tract, and numerous ligaments connecting inside of body wall to the digestive organ can be observed in the coelomic cavity. At beginning of the breeding season, two pairs of nephridia (1.5 cm × 0.3 cm) and storage sacs appear beneath the proboscis. During the higher breeding season, the storage sacs develop prominently (6-8 cm × 0.4 cm) and occupy the entire body cavity. The male and female sex cells are found not only in each storage sac but also in the body cavity mixed with coelomocytes and coelomic fluid.

1-4-1. Coelomocytes and body fluid

Except for the digestive tract, the body cavity is filled with two components out of the breeding season: the cellular component and non-cellular coelomic fluid. Coelomocytes are unique cells found in the coelomic cavity (Choe *et al.*, 1988; Jeong *et al.*, 1989; Kwon *et al.*, 2000). The low speed centrifuges (300 G) separate the coelomocyte from the coelomic fluid approximately by half in amount. Macrophages are infrequently found in the coelomic fluid and observed when the worms are infected (Miller, 1973; Garey & Davis, 1978; Masui, *et al.*, 1984; Frank *et al.*, 1985) Because of this, macrophages are not regarded as ordinary residents of coelomic fluid. During the breeding season, sex cells appear in the coelomic cavity intermixed with coelomocytes. All cellular components including coelomocytes, macrophages, and sex cells are not enclosed in compact tissue, but are found in loosely arranged tissue and floating freely in body fluid.

The term "coelomocyte" is used to describe single free somatic cells that ordinarily reside floating in the coelomic cavity. This name implies that these cells are phylogenetically "cells in the coelomic cavity" without indicating their function or origin. The coelomocyte is a "blood cell" responsible for respiratory metabolism (Terwilliger & Garlick, 1978), however it is reported that the "blood" is actually body fluid containing hemoglobin molecules (Ochi, 1969; Hall *et al.*, 1981; Magnum *et al,* 1983). On the other hand, coelomocytes are known to provide a substance called "urechrome" (Horowitz & Baumberger, 1941) necessary for developing oocytes. The term "urechrome" indicates that coelomocytes function as an accessory cell to the developing oocytes. It is likely that coelomocytes are multifunctional cells with important roles in hemoglobin synthesis, respiratory metabolism, and promoting sex cell development. Regarding their function as accessory cells, it has been noted that the number of coelomocytes varies greatly according to seasonal changes (Shin & Kwon, 1997). Cyclic changes of coelomocyte numbers are repetitive and coincided with each reproductive cycle. The number of coelomocytes reaches 100,000/mm^3 during the higher breeding season, and returns to normal levels (60,000/mm^3) during the non-breeding season. Parallel increases in the number of both sex cells and coelomocytes imply that coelomocytes participate in the reproductive cycle.

1-4-2. Digestive tract and respiratory organ

The *U. unicinctus* digestive tract extends from the base of the proboscis to the hindgut. The hollow digestive tract is folded many times lengthwise with a total length about 5-8 times longer than that of the body. The digestive tract and respiratory organ are not branched as seen in some other filter feeders (Remana *et al.*, 1975). However, the digestive tract can be functionally subdivided into three parts: the esophagus that is involved in food filtration (Shin & Lee, 1998), the midgut responsible for digestion, and hindgut that serves as a respiratory organ (Austin & White, 1981). Tissue organization of each component differs according to function.

1) Esophagus (foregut) and food filtration

The esophagus is dull white in color and about 12 cm long. This hollow organ is slender with a smooth surface and connects the proboscis to the intestinal tract. The entire inside wall of the esophagus is lined with ciliated epithelial cells. There are three kinds of cells: the ciliated cells, mucous cell, and other types of secretory cell containing lipid droplets. The ciliated epithelial cells (16 × 50 μm) bear numerous kinocilia extending toward the esophageal lumen (Shin & Lee, 1998). Hence, numerous kinocilia arranged along the inside of the esophagus remove or filter food materials floating in sea water. LM and EM studies found food material including plankton, organic debris, microorganism features, and non-organism particles between the esophageal kinocilia (Shin & Lee, 1998). Filtered food materials are transported towards the esophageal orifice or constrict and finally swallowed or further transported into the intestinal tract.

The flow of sea water into or from the esophagus is caused by negative and positive pressure created by muscular relaxation and contraction

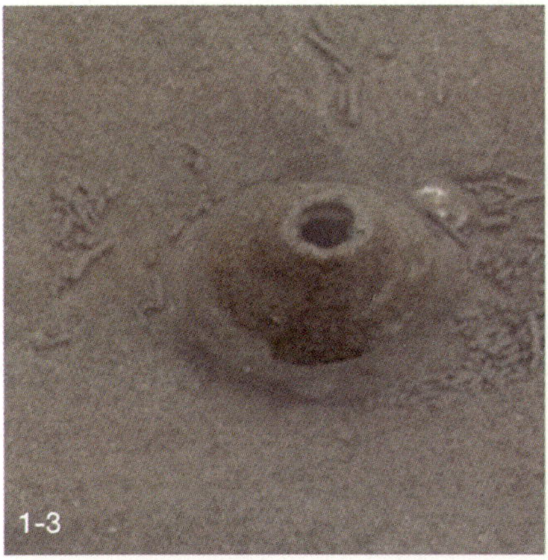

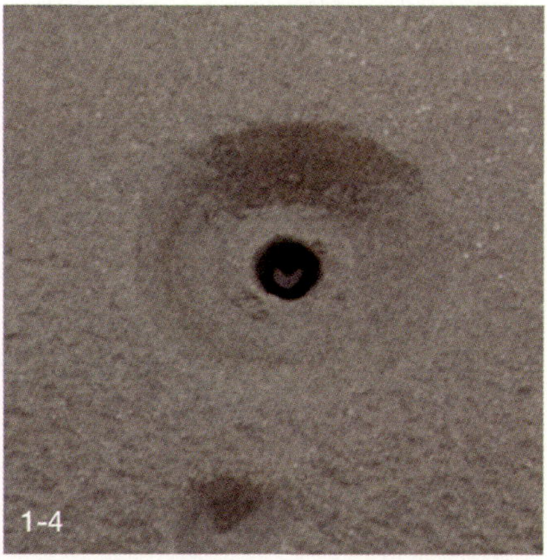

Fig. 1-3 Efflux pore of worm is convex and small bar-like feces of pale black in color are scattered around the efflux pore. Convex shape of the efflux pore is constructed by sticky sand-mud soil mixed with mucous secretions of worm.

Fig. 1-4 Double concavities of afflux pores are not yet completed at this time of season. The concavity of afflux pore is formed by rapid inflowing of sea water followed by mild out flowing of sea water including sand particles and debris.

Fig.1-5 A set of efflux and afflux pores on the habitable tideland.One line distance between efflux and afflux pore measure 1~1.5 m average, but sometimes, the pores are approached 60-70 cm as seen in the figure.

(Muneoka & Kamura, 1982; Pritchard & White, 1981). Negative pressure due to muscular expansion is transmitted to the esophagus by numerous ligaments that directly connect the body muscle to the esophagus. The esophageal volume increases about 10 times the usual volume.

2) Intestine

The intestinal tract can be easily recognized given its rough surface and coloration, and is subdivided into two parts. The frontal gut is red and the midgut has scattered patterns of red and pale black. The red color of the digestive tract may come from the digestion of plankton while the black color represents a final metabolite of the food including mud from the tideland. Between the midgut and hindgut, a stenosis that functions as an anus can be observed. The semi-solid excretions of the worm are blackish in color and rod-like (Fig. 1-3). They are packed within a fine meshwork about 3 mm x 1 mm in size. The feces are first excreted into the hindgut that functions as the respiratory organ and widen the connection to the midgut. During the outflow of seawater from the hindgut, the feces are expelled from the worm onto the tideland surface.

3) Hindgut: a respiratory organ

The hindgut is a semi-transparent, grey and hollow organ consisted with cellular meshwork. It is slightly shorter than half of the body length and two about three times wider than the intestine. The wall of the hind gut is composed of layers of epithelial cells that form a thin membranous structure with different textures, and can be expanded more than five times its original volume (Austin & White, 1981). A stenosis that serves as an anus appears as a demarcation line between the intestine and hindgut. Numerous ligaments connect the body wall to the surface of the hindgut.

The negative pressure due to the partial expansion of the body muscle is transmitted through the ligaments onto the hindgut. This leads to synchronous expansion of the hindgut as well as an inflow of seawater. The expansion exposes a thin membrane comprised of extracellular materials between the cellular components of the hindgut. Through the membrane, oxygen rich seawater and body fluid come into close contact and gas exchange may occur between the sea water and the body fluids that contain hemoglobin molecules (Hoffman & Magnum, 1970; Garey & Riggs, 1984). In *U. caupo*, pO_2 and pCO_2 are measured prior the inflow and after the outflow of seawater to evaluate changes of pO_2 (Austin & White, 1981).

The inflow of sea water is briefly retained in the hindgut by a strong contraction of the tail muscle. The deoxygenated sea water is then pumped out of the hindgut and the efflux pore. When positive pressure by the body muscle is applied to the hindgut, sea water is ejected in a high water column that travels above the tideland surface. Under experimental conditions, water is expelled in jets 30~40 cm high. The expulsion of sea water also refreshes the stagnant water of the burrow reservoir.

4) Nephridia

Similar to the gonads, nephridia positioned beneath the proboscis undergo structural fluctuations according to the seasonal cycle. Out of the breeding season, it is difficult to see the nephridia since they are reduced in size. At beginning of the breeding season, two pairs of nephridia appear as thread-like structures attached to the bases of proboscis, and begin to grow and extend backwards toward the worm. Each nephridium is a slender sac-like structure with an undulate or rough surface and is semi-transparent white in color. The size of the nephridia varies slightly; some of them reach 5 mm × 4-5 cm in size during the higher breeding season. The functions of the nephridia are not yet known but may be associated with the sexual cycle. The shapes of the nephridia do not vary according to gender. The storage sacs and nephridia differ in color, size, and tissue organization.

1-5 Storage sacs

In the literature, various names are used for the storage sacs of the worm. One reason may come from the considerable structural differences of the sac-like structures in which the sex cells accumulate during the higher breeding season. These sac-like structures have been called "gonoducts" (Sato, 1939), "storage sacs" (Sawada & Ochi, 1975; Paul, 1975; Gould-Somero, 1975), and or sex organ (Jeong *et al,* 1989). The reason for the disagreement about the name for the sex organs is based on the fact that gametogenesis occurs in the coelomic cavity. Various developmental stages of oogenesis and spermatogenesis as well as mature gametes are observed in coelomic fluid. In the coelomic cavity, the gametogenesis is associated closely with the changes in the number of coelomocytes coincides with every reproductive cycle (Shin & Kwon, 1997).

Gould-Somero & Hollander (1975b) reported that mature coelomic gametes accumulate in the storage sacs. How the mature gametes are gathered and accumulated in the storage sac is not currently understood. Recently, a pore or an opening was observed at the distal tip of each storage sac, and it was suggested that the pore is involved gamete storage (Shin & Kim, 2002). Another source of the term "storage sac" may come from the fact there is no spawning pore that connects the coelomic cavity directly to outside of the worm. Hence, it is apparent that coelomic gametes are stored in the storage sac before being discharged. On the other hand, there is some evidence that male gametes are produced in the storage sac (Jeong *et al.*, 1989; Shin & Kim, 2002). The pathway of gametes and sex organs of *Urechis* should be examined to better understand *Urechis* biology.

1-5-1. Male storage sac

As like the female storage sacs, the male storage sac (MSS) consists of two pairs of slender sac-like structures (4~6 cm in length) filled primarily with mature male gametes prior to the mating season. During the beginning of the higher breeding season, gamete accumulation causes expansion of the upper proximal portion and turns this portion white while the lower distal portion is remained in slender and dark brown in color. The developing male gametes can be found in the interim portion between the proximal and distal portions. The male gametes are enclosed in groups of sperm balls containing somatic cells and a number of male gametes (Jeong *et al.,* 1989; Shin & Lee, 1998; Shin & Kim, 2002). The expansion continues until the entire MSS is filled with mature sperm. During the higher breeding season, mature sperm is discharged from the MSS through the sexual pore that opens laterally below the proboscis.

According to Jeong *et al.*(1989), two different types of spermatogenesis simultaneously occur in the coelomic cavity in male *Urechis*. One type takes place within the organized tissue. The other type occurs within coelomic fluids but not distinctive or organized tissues, similar to some polychaete (Fischer, 1975; Fischer & Dhainaut, 1985). The former process is known as localized spermatogenesis while the latter is called diffuse spermatogenesis (Korschelt & Heider, 1902).

Recently, it was observed that storage sacs are not blind sacs but a pore or opening is found at the distal tip of each MSS (Shin & Kim, 2002). It can be inferred that mature sperm are selectively gathered through the distal pores and accumulate in the storage sacs. However, the selection and storage mechanisms of coelomic mature sperm have not been characterized. The presence of spirally arranged fibrous tissues in the distal pore (Shin & Kim, 2002) and peristaltic motions of the body muscle may play a role in conducting coelomic sperm into the storage sac.

1-5-2. Female storage sac

Out of breeding season, the female storage sac (FSS) cannot be found in the coelomic cavity. At beginning of the season, the two pairs of FSS appear beneath the proboscis and begin to fill with mature female gametes. The FSS is yellow color due to yolk that is deposited into the female gametes. Except for the color, the features of the FSS are similar to those of the MSS. Mature female gametes are primarily found in the FSS while developing gametes are located in the coelomic cavity. The authors are of the opinion that the oogenesis continues diffusely in the coelomic cavity and mature gametes are stored selectively in the FSS during a final developmental stage (Gould-Somero, 1975; Paul, 1975; Shin *et al*., 2000). In reference of the male gametes developed in the male storage sacs (Jeong *et al.,* 1989; Shin & Lee, 1998), it is not yet conclusive whether the coelomic cavity is the proper site of the oogenesis. Korschelt and Heider (1902) report that once a pattern of sex cell production is determined in a species, the same patterns progress in both genders of the species. In assumption of the

MSS is the site of spermatogenesis (Jeong *et al.,* 1989; Shin & Lee, 1998; Shin & Kim, 2002), the description may not concur with gametogenesis in *Urechis,* since both localized and diffuse spermatogenesis is observed in males. Female gametes stored in the FSS are apparently mature oocytes capable of fertilization. The mature oocyte prepared from the FSS or coelomic fluid is in fact a premeiotic gonial cell because the oocytes are fertilized during a premeiotic phase (Kwon *et al.,* 2000; Kwon *et al.*, 2003). As above, mature female gametes are also prepared in the coelomic cavity. The oocytes of both derivations are 130 μm in diameter, yellow in color because of the deposited yolk, and have cytoplasmic indentations as well as nuclear invaginations, same in fertilization capacity.

The oocytes of *U. unicicntus* are used for studies in various fields including biochemistry, cell biology (Das, 1976; Miller, 1973; Frank & Davis, 1983; Frank *et al.,* 1985), and developmental biology. These oocytes have two characteristics that make them suitable for investigations. First, the oocyte is a single free cell and its growth progresses solitarily in coelomic fluid. The development and differentiation of the oocyte occurs without the presence of an auxiliary somatic cell (Gould-Somero & Holland, 1975a; Yamamoto *et al.* 1973). Second, the growth of the oocytes is not synchronous so that oocytes at various developmental stages can be prepared in coelomic fluid. The developmental stages can be identified simply by the amount of yolk deposition and appearances of various cellular architectures. On the other hand, mature oocytes are homogeneous in size and fertility, and selectively stored in the FSS. Semi-transparency of oocytes permits a view of the oocyte interior including germinal vesicle, nucleolus and some cytoplasmic components.

1-5-3. Spawning and fertilization

During the higher breeding season, the worm habitats are no more than a single line of spiral "U"-shaped burrows. The worms connect their own burrow with others. Females and males congregate in burrow water reservoirs. Male and female sex cells are discharged through two pairs of sexual pores and the gametes are then fertilized. The fertilized oocytes settle to the bottom of the stagnant water in the reservoir. These develop into ciliated trocophore within 36 hours and are then pumped out from the water reservoir by the cyclic afflux and efflux of the worm's respiratory organ or trocophores rise from the burrow to the surface of sea water at higher tides.

The ciliate trocophores travel on waves, are distributed over the tideland, and then permeate and submerge into the mud-sand of the tidelands. The trochophores begin to assemble burrows using ciliary movement. During the development of trocophores into adults, the worms reside alone in their burrow. In July and August just after the breeding season, young worms resembling white maggot can be found 3~5 cm below the surface. It is interesting that adult worms are concealed deep underground while young larvae are found not far from the soil surface.

1-6 Uses of *U. unicinctus*

U. unicinctus is an economically and ecologically important species along the whole Korean coast. Mongsan-po beach is one of the largest and famous worm-producing districts. Sato (1939) investigated the distribution and use of this worm in South Korea. In the south coast of Korea, *Urechis* are used as live bait or dried food. However, the worms have been neglected by people living on the west coast until the 1970s. From the 1970s to 1980s, a baseless rumor was wide spread along the west coast stating that the worms were an effective aphrodisiac and used as traditional medicine by some people. The rumor seemed to be based on the external features and color of the worm.

In the mid-1980s, the worms became a source of fixed income for the residence around the Mongsan-po coast. From February to mid-April, fishermen and women caught the worms. Over time, the worms are overhunted. According to a wholesaler in the 1980s, an average of 5 million worms was caught every winter and consumed in nearby cities and Seoul. At that time, fish farmers had a special interest in the worms and intended to cultivate the worms. In 1984, one worm costs 20 Korean Won. Presently, one worm costs between 1,200-1,500 Korean Won, an increase of 60-75-folds.

The worm decreased dramatically in number up to the end of the 1980s. During this time happen all at the same time, people drove cars across the coast, and the beach was used as a runway for light airplanes. Environmental management began in the middle of 1990. Consequently, worm populations began to increase, and the habitat conditions seemed to be restored simply by prohibiting the movement of automobiles and airplanes along the coast.

Although the successful environmental management with voluntary cooperation of people helped worm conservation, it seems that the worms can protect themselves. Fortunately, the worms cannot be caught and consumed after mid-April. Sex cells of both genders begin to accumulate and rapidly fill the storage sac. The sex cells and organs give off an extremely unpleasant odor, making the worms unfit for human consumption. This may be a key to worm survival. Additionally, it is impossible to catch the worms when the tide rises, at which time spawning occurs. It is somewhat ironic that the worms, which are often consumed as an aphrodisiac, become undesirable as they themselves enter the reproductive phase of their lives.

1-7 Materials and methods

The subject matter of the monograph is description of single species, *Urechis unicinctus* collected from the intertidal zone on the west beaches of South Korea and are reared as described by Shin *et al.* (2000, 2005a, 2005b). Although the practical and essential methods applied to every experiment are differed from each other, the methods of material handling or treatments are not greatly biased with each of the experiments. To avoid the repetitive descriptions of the resembled methods, the basic methods of the material handling are synthesized and summarized, the essential methods are narrated separately.

The worms, *Urechis unicinctus*, are collected during the breeding season, from end of January to beginning of May. Since the worms live normally deep underground of the tidelands, it is not easy, still not unattainable to gather the worm in the breeding season. In the higher breeding season, the worms become more active, crawl about the burrow on the tideland surface beneath, easier to capture it for skilled fisherman. The worms living in pulverized sand soil cannot be easily captured by the reason of the mixture of much sand and seawater flows simply inward of the excavated burrow.

The worms are cultivated in laboratory under controlled conditions for a few days, but it is hard to maintain a long term culture. The worms are not on the feed, and for a while of the worms are off its feeds, anatomy reveals often the deteriorations of the visceral organs, even though the worms maintain the proper external shapes. The prevention of infection is conspicuous to cultivate the worm. Every empty hand touch causes coming to serious fester at the contact site of the fingers. The immersing the worm in a diluted solution of dichromate salt makes the fester better, but the effect is tentative, cannot heal the fester thoroughly. The worm is kept at about 16~18℃, used normally within 48 hours after the capture. The special attentions are given on the transportation to avoid unbearable swinging and rocking broken out of automobile, which affected greatly the vitality of the worms.

The contents of the coelomic cavity, coelomic fluid and cellular components including coelomocytes and gametes of both genders are prepared by longitudinal opening of the body wall. Making small aperture through the body wall is not effective or useless to get one or more components of the coelomic contents. The worm stops up the aperture tightly with parts of intestine every time of the preparations continued for 30 years. The stopping is not permissive to leak a drop of the coelomic fluid and other constituent of the coelomic components through the bored aperture. It is likely that the close up of the aperture by the visceral organ are not chance events, but active intensions of the worm. The needle insertions come to the same results at every time as making an aperture on the body wall. The longitudinal opening is necessary to prepare the coelomic contents ensuring an amount.

As is described previously, the gametes exhibit seasonal fluctuations, repeat in appearances and disappearances according to seasonal changes. The seasonal changes can be taken a matter into account to prepare the female and male gametes. On the other hand, the coelomocyte is the ordinary resident in the coelomic cavity, can be prepared all the year round except the seasonal interval from June to end of September. The worms are hidden itself in deep underground, disappeared completely without trace from the earth surface during the seasonal interval. Recently, the worm become an item on sale, some fishermen supply the cultivated worms to wholesaler out of the season.

1-7-1. Physical conditions of the material preparations

The mature gametes of *Urechis* are reliable materials to demonstrate all kinds of fertilization events visualized in real time scanning. This is because of the semi-transparent oocyte, the entire processes of spermatozoon infiltration or penetration and the various corresponding responses of the oocyte can be continuously observed with little bit of fortunes under microscope. On the assumption of encountered sperm and oocytes, one of the essential conditions of successful fertilization is that the chance of the fertilized oocytes to meet favorable environment for further developments. In reality, the conditions of the habitable tideland are inconstant and extremely changeable. The salinity concentrations for instance are changeable at every moments of time by intermixing rainwater and sea water. The temperatures in the breeding season seesaw in range from 8~9℃ to 25℃ at sunny days. As like many other animals living in the tidelands, the worm seem well adapted to the environments created specially at the spawning season. A favorable condition enables the artificial fertilization and further developments are described as follows.

1) Temperatures

The gametes, especially the oocytes whether being fertilized or not, are sensitive to changes of temperatures. The plasmolyses of the unfertilized oocytes are frequently observed, and number of fertilized oocytes develop abnormally at a higher temperature above 20℃. On the other hand, a lower temperature delay the developments considerably more than expected. The delay for a little while seems not seriously affected the developments of the fertilized oocyte, but inconvenient to find out exact time scale of the normal developments. By the reasons, efforts are given to select an adequate temperature.

An acceptable temperature, e.g. the highest low temperature of sea water used in experiments ranges at about 16~18℃, differed from the sea water temperature of about 15℃. It is likely that the temperature of the natural brackish water (NBW) stagnated in the water reservoir is slightly higher than the temperature of the oceanic sea water flowing over the tideland surface. In addition, at a higher temperature than 18℃, parasitic rotatorians begin to activate to destroy much number of oocytes for furniture of foods. The rotifers are internal parasites reside in the coelomic cavity, approach closely the oocyte rotating circular cilia ceaseless. The ciliary movements bore opening through the oolemma, through which the ooplasm flow out of the oocyte. The rotifers provides the ooplasm with foods. Below the temperature 18℃, the activity of the rotifers are tolerably decreased for oocyte.

2) Salinity

The other critical factor to be considered for the gamete treatments is the quality of the sea water. The worms live in tideland where rise and fall of sea water are continuously repeated. And the breeding season of the worm is corresponded to the wet monsoon at which the tideland is frequently exposed to heavy rain. The fresh water from the hillock flow over the tideland surface infiltrating into the soil and mixed together with sea water from the ocean making kind of brackish water. The salinity concentration of the brackish water at the higher breeding season is maintained between 18~2 2‰. To meet favorable conditions for fertilization, we used natural brackish water (NBW, pH 8.1). At the low salinity of the NBW, the indentations of oocytes are recovered round shape exposing the most of microvillous tips, the sperm receptors, on the round contour line from the concealments of the indentation.

Gould-Somero(1975) prepared oocytes from the storage sac of *U. caupo*, described the concave structures may be indications of maturity of the oocyte. In a salinity concentration over 25‰, the concavity of the *U. unicinctus* oocyte begin to be appeared again, it is hard to deal with the materials and carrying out exactly timed experiments. Using 18~22‰ of NBW salinity, it is able to avoid over swelling or deflation of the oocytes, and induce higher fertilization rate (~80%).

Conventionally, the 0.45μm millipore filtered NBW is always used in experiments. The filtrations remove much more amount of organic and inorganic debris inclusive of parasitic rotifers. The organic debris is sticky on the oocyte and disturbed frequently the observation sights.

1-7-2. Gamete preparations

1) Preparations of mature sperm and developing male gametes

The mature sperm of the worm are prepared from the storage sac and coelomic fluid by longitudinal opening of the body wall. In the breeding season, the mature sperm accumulated in the two pairs of male storage sacs, can be prepared whether from the storage sacs or from the coelomic fluid. It is preferred except the special cases to use mature sperm prepared from the storage sac, because of its homogeneous maturity and fertilization capability, not demanded forcibly extra separation steps and methods. The storage sacs of both genders are delicate cellular membrane, easily broken down allowing flow out of the gametes. To prevent the flowing out, the proximal end of the storage sacs can be tied with hair prior to cut out the storage sacs.

The mature sperm freshly prepared from the storage sacs are embedded and entangled in highly viscous fluid showing no indication of movements and activity. But skimmed with sea water, the sperm begin random vigorous activities which continue for about one and half minutes. Thereafter, the activities decrease rapidly and empty activities continue until 3 min. It is likely the fertilization capability of sperm is durable for the first half period of the life time. In the artificial fertilization experiments, the viscous fluid of the entangled sperm takes time to be dispersed in the suspension of oocyte, it is difficult and inconvenient to determine the correct fertilization timing or to ensure balanced fertilization effects. As a matter of convenience, the viscous fluid of the sperm is previously dispersed rapidly in a small amount of NBW.

While the mature sperm are stored in the storage sacs, as above, the developing male gametes are found in coelomic fluid. Since the developing male gametes share the coelomic fluid with coelomocytes, it is necessary to isolate the male gametes separately from the coelomocytes. Brief centrifuge (~300 rpm) of the coelomic fluid (~15 ml in maximal amount) separates approximately the upper supernatant of the male gametes from the precipitate containing coelomocytes. Further divisions of the upper layer are not obvious to naked eyes in practical experiences, but lower supernatant contains developing gametes and top layer includes mainly mature sperm, while the spermatid cluster and spermatid are found in the middle layer.

The developments of the male gametes are asynchronous. The preparations of the coelomic fluid at any arbitrary seasonal time exhibit in principle all of the developmental stages of the male gametes. The preparations at an earlier seasonal time distantly far from the spawning season, for instance at the end of February, expose coelomic fluid containing mature sperm. However, for the preparation assuring exactly timed stages and amount of the materials, it is preferred to select the adequate seasonal time corresponding the developmental stages. The mature sperm can be prepared after end of April, and developing male gametes being separated from the sperm cluster are found in the preparations from mid-March to mid-April, and young male gametes belonged to sperm cluster are provided from the worm prepared mainly in February.

2) The preparations of mature oocyte and developing female gametes

The female gametes can be prepared in principle with the same methods used in preparation of the male gametes. But the peculiarity of the female gametes distinguished itself in details from the separation processes of the male gametes.

The mature oocytes are prepared whether from two pairs of storage sacs or from the coelomic fluid, which contain also developing and mature female gametes. It is preferred preparing mature oocyte from the female storage sac, because of its homogeneous maturity and fertilization capacity, not necessitated further separation procedurcs. As is described in section of the salinity control, the mature oocytes show cytoplasmic indentation and nuclear invagination. The concavities especially of the cytoplasm are spontaneously restored round shape in NBW lower than 22‰ in salinity. The mature oocyte of the storage sacs are surrounded by a little amount of viscous fluid. To remove the viscous fluid and induce restoration of the concavities, the oocyte is washed in 0.45 μm Millipore filtered NBW of salinity 18~22‰. The concavities of the oocyte are restored in the NBW.

The developing oocytes are prepared from the coelomic fluid as described above. It is necessary to select the mature oocytes from the coelomic fluid to achieve higher fertilization efficiency, since the growth and developments of oocytes are asynchronous, and the coelomic fluid contains female gametes of various developmental stages. The coelomic fluid is centrifuged (~300 rpm) briefly, and the supernatant is separated from the precipitated coelomocyte. Using Nylon mesh of 100 μm pore size, the supernatant including oocytes of various developmental stages is filtered to select the mature oocytes. The most of mature oocytes (above 100 μm in diameter) are filtrated and isolated on the mesh. Because of the flexibilities of the oocytes, the most of the oocyte of 100 μm in size is infiltrated through the pores.

The oocytes prepared from the coelomic fluid and storage sacs are equivalent in fertility developing into larval stage. A various meiotic events e.g., formations of meiotic apparatuses and asters, extrusions of two polar bodies and pronuclei fusion have been observed during developments of the fertilized oocytes of both origins. The results indicate the oocytes of both derivations are germinal vesicles encountered the meioses.

1-7-3. Artificial fertilization and parthenogenetic activation

1) Artificial fertilization

The oocytes and sperm are artificially inseminated to observe the fertilization events and its subsequent features of the developments. The oocytes of both derivations are used directly after the brief washing. The oocytes are settled down spontaneously on bottom of the NBW and buffer solution, centrifuge is not necessitated. A little number of the floating oocytes is discarded.

The oocytes are washed and suspended in NBW. The amount of oocytes and NBW are maintained in the rate of 1:100(v/v), and the sperm and oocytes in the ratios of 0.1 : 1(v/v). The artificial fertilizations are fulfilled immersing the oocyte suspension into the sperm suspension or conversely intermix the small amount of sperm suspension into the oocyte suspension. In practical sense, the former is more convenient by the reason of the additional agitation is not necessitated. The pouring of the large amount of oocyte suspension into the small amount of sperm suspension is accompanied itself by enough agitation of the mixed suspension.

The vigorous activities of the sperm are not kept up above 2 min and about 3 minutes beyond encountering sea water, the most of sperm are inactivated and continued empty activities, settled down on the bottom. After 1 min of the intermixing, the excessive sperm are washed out several times with the NSW. As above, the temperature and quality of the NBW are critical conditions for the appearances of patterned fertilization events and subsequent normal developments.

To observe more detailed initial processes of the sperm-oocyte interactions, it is sometimes necessitated to slow down the processing rates of the sperm-oocyte interactions. For the purpose, the oocytes are prefixed in prefixative (*See* Section 7). After washing, the sperm suspension is added into the appropriate amount of the prefixed oocyte suspension. The sperm-oocyte interactions are eventuated on the prefixed oocyte, and the subsequent fixation procedures fix and preserve the prefixation effects making it possible to observe much more structural details of the interactions. The fixation effects of the prefixative solution are much slower to fix the small area of the sperm receptors of the microvillous tips, or not enough fixing the oocyte completely. On the other hand, once the oocytes are immersed in the glutaraldehyde of concentration above 1%, the sperm are not dared to come near the fixed oocyte.

2) Parthenogenetic activation of the oocyte

The artificial fertilizations are paralleled with parthenogenesis to evaluate the timing and actual states of the parthenogenetic events in contrast to the normal fertilization. The parthenogenesis are induced immersing the oocytes into sea water of pH 8.6~8.7. Since the pHs of the NSW are maintained normally 8.1~8.2, the pH of the NSW are adjusted with additions of ammonium acetate (Sigma). The parthenogenetic oocytes are precipitated on the bottom of the NSW, and the suspensions are continuously and mildly aerated, so that the fertilized oocytes on the bottom are not seriously agitated. The developmental time scale is similar with the artificial fertilization, the parthenogenesis induce the raise of the fertilization envelope within 1 min and occurrence of GVBD at 2~3 min. After about 1 min, the surplus sperm and parthenogenetic agents are washed out several times using the same NBW. The fertilization and phathenogenetic events of the living materials are observed under Domarsky interference microscope (DIC) (Diaphot, Nikon).

1-7-4. Electron microscopy

1) Transmission and scanning electron microscopy (TEM, SEM)

It is repeatedly reported in literatures that marine organism, especially the extracellular layer of fertilized oocyte, are hard materials to be fixed for electron microscopy (EM) (Elinson & Manes, 1978; Gould-Somero, & Holland, 1975a). Moreover, in proceeding of the experiments, the extracellular layer takes on importance for fertilization and maintaining all of fertilization events including the subsequent developments. The importance of the extracellular layer is not next to oocyte itself, should not be underestimated to understand the entire processes of fertilization. When the extracellular layer is removed previously, the oocyte cannot be fertilized normally, nevertheless the further development of the fertilized oocyte is not progressed normally (*See* Parthenogenesis). Within the ranges of fixatives examined, the following triple fixation method preserved the fine structural details and serial changes of the developments.

The materials are prefixed in 0.1M phosphate buffered (pH 6.8) 4% formaldehyde containing 0.5% glutaraldehyde for 1~4 hs at 4℃. The prefixation solution maintains roundness of the fertilized and pathenogenetic oocytes, and sperm-oocyte interactions are structurally stable during the entire processes of the subsequent preparations. The latter is a remarkable effect of the fixative. Using other fixatives, it is not impossible to observe the details of the sperm-oocyte interactions on the oocyte surface, but the chances are greatly decreased. The prefixation solution preserved the sperm receptor on the microvillous tips. When the oocytes are treated previously in the prefixation solution, the prefixed oocytes are exposed in sperm suspension, the oocyte-sperm interactions are occurred, even thought the subsequent fertilization events of the oocyte are not progressed further. The antigenic preservation of the prefixation solution is used in immunoelectron microscopy.

The materials are subsequently fixed in 0.1M phosphate buffered 2.5% gluaraldehyde for 2 hours, washed in distilled water, postfixed in 1% OsO_4 for 2 hours. After washing in distilled water, the materials are dehydrated in ascending series of graded ethanol and prophylene oxide, embedded in Araldite mixture. The blocks are incubated for overnight at 40℃, 60℃, and 65℃, respectively. Centrifugation is necessitated once for all prior to prefixation, because of the rapid precipitation of the materials in each step. Blocks are sectioned on OM-2 Reichert ultramicrotome. The semithin section are stained in Richardson's solution (1960) at 60~80℃ and thin sections are contrasted in saturated uranyl acetate and lead citrate. The sections are observed under EM, JEM 1010B, JEOL.

In study of the gametogenesis, attentions are paid on the coelomocyte playing a role in functional supports of the developing gametes. One of the reasons is the gametogenesis of both genders is diffuse gametogenesis. The singular gametes are developed and differentiated floating in the coelomic fluid without associating accessory cells of somatic origin. The other is that the coelomocyte is the unique cellular components assigned the coelomic cavity to the developing gametes. Moreover, the analyses reveal the compositions of the coelomic fluid varied in accord with seasonal changes. Since the coelomocyte is known as blood cell (*See* coelomouyte), it is likely that the coelomocyte is a kind of multifunctional cell, not confined within a respiratory function, especially in the breeding season. Because the coelomocyte is single cell, not confined itself in a compact tissue, it is regarded that the multifunction is greatly related with a structural changes of the coelomocyte surface. To observe the seasonal changes and differences of the coelomocyte surface, the coelomocytes are prepared into seasonal groups and observed under SEM.

The brief centrifuges (~300 rpm) separate the coelomocyte from supernatant coelomic fluid. The coelomocytes are washed three times in 0.1M phosphate buffer (pH 7.4), and fixed in 2% glutaraldehyde for two hours. After washing with the same buffer, the samples are postfixed in 1% OsO_4 for two hours. The samples are dehydrated in alcohol series, substituted in ascending grades of isoamyl acetate, dried in critical point dryer (CPD), and coated with Au evaporation 150 Å in thickness. The samples are observed on scanning electron microscope (JSM-5410LV, JEOL).

2) Anti-actin immunogold particle (AAIP) labeling and EM observation

To observe the spermiogenesis, the spermatids isolated from the coelomic fluid are labeled with immunogold particles and the labeled spermatids are brought onto EM. The coelomic fluid of male are briefly centrifuged to collect the developing male gametes and washed in PBSB (0.1 M phosphate buffer, 0.15 M saline solution, 1% solution of inactivated BSA, pH 7.2) (Slot & Geuze, 1989). The AAIP (10 nm) labeling and electron microscopy used in this work are in principle followed De May's "postembedding methods" (De May, 1983). For the invertebrate materials, some details are modified as described below. The materials of immunogold labeling groups, the anti-actin and anti-tubulin groups are fixed in 0.1 M phosphate buffered 4% paraformaldehyde with 0.5% glutaraldehyde containing 0.1% $CaCl_2$ (Gould-Somero & Holland, 1975a; Elinson and Manes, 1978). After the fixation, the materials are dehydrated and embedded in Araldite mixture in accord to the same methods described above. The blocks are sectioned and the 70-80 nm ultrathin sections are gathered on nickel grids, etched in 10% H_2O_2 for 3-5 min. Thereafter the sample grids are treated with gelatin, glycine solution and 1% BSA solution respectively, rinsed 3 min in the PBSB solution. The samples are incubated for 4-6 hours with amoeba monoclonal anti-actin (mouse–IgG isotype, Sigma) (1 : 100~200) in moisture chamber at 4℃. The gold particles (10 nm) conjugated goat anti-mouse-actin is diluted 200 times, and the dilutions are applied to mark the primary antibody. The anti-actin immunogold particles (AAIPs) incorporated samples are then fixed in 1% glutaraldehyde for 1 min, and after the washing in the same buffer the samples are postfixed in 0.2% OsO_4 for 1 min. The sections are contrasted in saturated aqua solution of uranyl acetate and lead citrate, and observed on JEOL 1010B transmission electron microscope.

On the other hand, for the observation of the tubulin residues, the anti-tubulin groups are treated with the same methods used in actin labeling experiments. After the etching in 10% H_2O_2 for 3-5 min and BSA treatments, the samples are washed in the PBSB. The primary Ab, monoclonal anti-α-tubulin (mouse IgG isotype, Sigma) is diluted 1,000~2,000 times with the PBSB, and the sections are incubated in the Ab solution for 1 hour. After washings, the sections are incubated in the gold particle conjugated secondary Ab solution, goat anti-mouse IgG-gold particle conjugate. The concentration of the immunogold particles (10 nm) is adjusted to be 5.85×10^{10}/ ml, and the sections are incubated in the secondary Ab solution for 1 hour in moisture chamber. The Ag-Ab incorporation are fixed in 1% glutaraldehyde and 0.2% OsO_4, and sectioned as described above.

1-7-5. Immunofluorescence methods and confocal laser scanning microscopy

1) FITC-Phalloidin staining of the male gametes

To make the polymerized actin visible, the commercial FITC-phalloidin is dissolved in methanol to make 3.3 μM FITC-phalloidin solution. This solution is diluted 50-100 times with 0.1 M Tris buffer just before use. The male gametes are smeared gently on cover glasses which have been previously treated with 10% aqua solution of Poly-L-lysine. The smeared sperm are covered with the FITC Phalloidin for 1 hour in moisture chamber at 4℃, and the excess FITC-phalloidin solutions are washed off with the same buffer. The cover glasses are mounted upside down on slide glass using Gel mount.

2) Anti-actin-FITC staining of the male gametes

For observation of the depolymerized actin, the commercial amoeba monoclonal anti-actin conjugate is diluted with 0.1 M Tris buffer (1: 1,000). The prefixed male gametes are treated in anti-actin solution to imprint actin monomer for 5~6 hours in moisture chamber. The antibodies are then marked with the anti-mouse IgG-FITC conjugates diluted in the same buffer. The preparations are observed by CLSM using 15 mW Krypton/ Argon laser power (15-30%).

3) Anti-tubulin-FITC staining of the male gametes

Since the FITC-phalloidin and anti-actin-FITC conjugate showed weak reaction to flagella. It is often obligatory to view the whole sperm structures under CLSM. The prefixed male gametes are treated in anti-α-tubulin solution to imprint α-tubulin. The antibody solutions are diluted 1,000 times with 0.1M Tris buffer. The antibodies are then marked with the anti-mouse IgG-FITC conjugates diluted in the same buffer. The immunostaining methods of multisteps, FITC-Phalloidin and anti-actin-FITC followed by anti-α-tubulin are more effective to observe the imprinted immunofluorescence on the male gametes.

4) Incorporation of anti-tubulin-FITC into the fertilized oocytes

To observe microtubules, microtubule associating structures and meiotic apparatuses (Maps), the sperm and fertilized oocytes are prepared according to methods of confocal laser scanning microscopy (CLSM). The unfertilized oocytes are prepared with the same methods to contrast the experimental groups. The materials are fixed in cold methanol (-20℃) for 24 hrs, and then is washed three times in modified phosphate buffered saline solution (mPBS; NaCl 8G, KCl 0.2G, K_2HPO_4 0.99G, KH_2PO_4 0.2G, pH 6.8).

To observe meioses, polar body formation and the fertilized spermatozoon taking residence in the oocyte cortex, the fertilized oocyte are treated in 2% Triton X-100 for 12 hrs to enhance the reaction efficiency to anti-tubulin, and then washed three times in the same PBS. The primary Ab, the anti-mouse β-tubulin are diluted in PBS containing 0.1% Triton X-100 in ratio 1:200, and the materials are incubated in the primary Ab solution for 1 hour at 37℃ in moisture chamber. After washing with PBS, the incorporated primary Ab are marked with the FITC-conjugated goat anti-mouse IgG (Sigma) diluted 200 times for 1 hr at 37℃. The male and female pronuclei are stained with propodium iodide diluted in alcohol, embedded in glycer gel (Dako), and observed on CLSM (MRC-1024, Bio-Rad) using Kr-Ar dual laser.

1-7-6. Analysis of *Hox* genes

Adult specimens of the echiuroid *U. unicinctus* were collected from the intertidal zone of the west beaches of South Korea and were reared as described by Shin *et al.* (2005a, 2005b). Genomic DNA was isolated from body wall tissues of adult *U. unicinctus* containing epidermis and muscle layers to minimize DNA contamination by ingested prey and parasites. These tissues were homogenized with a Teflon-glass homogenizer in buffer A (100 mM Tris-HCl, pH 7.5; 100 mM EDTA, 100 mM NaCl, and 0.5% sodium dodecyl sulfate). The homogenate was incubated at 65℃ for 30 min and combined with a 6 M LiCl/5 M KCH_3COO solution (1:2.5, v:v). To extract the genomic DNA, a mixture of phenol/chloroform/isoamyl alcohol (25:24:1, v:v:v) was added. Finally, genomic DNA was precipitated by the addition of isopropanol and washed in 70% ethanol. The isolated DNA was resuspended in Tris-EDTA buffer.

Total RNA was prepared from 100 blastulae and 100 trochophore larvae using TRIzol reagent (Sigma, USA) according to the manufacturer's protocols. Poly $(A)^+$ mRNA was isolated using a Dynabeads mRNA DIRECT Kit (Dynal, Norway) following the manufacturer's protocols. Poly $(A)^+$ mRNA was quantitated spectrophotometrically and then stored at -85℃ until use. One microgram of poly $(A)^+$ mRNA derived from blastula and trochophore larvae was reverse-transcribed using an ImProm-II Reverse Transcription System (Promega, USA).

To search for *Hox* genes, we performed polymerase chain reaction (PCR) on genomic DNA or cDNA synthesized by reverse transcription of blastulae and trochophore larvae mRNA. PCR was performed in a reaction volume of 100 L containing 1 L template cDNA, 2.5 μM of a dNTP mixture, 2.5 mM $MgCl_2$, 2.5 pmol of each of primer, and 0.3 U Taq polymerase (TaKaRa, Japan). The PCR conditions were as follows: denaturation at 9 5℃ for 1 min, annealing at 50-55℃ (depending on the primers) for 1.5 min, elongation at 72℃ for 1 min, and a final extension at 72℃ for 10 min.

To amplify *Hox* gene fragments isolated from genomic DNA, we used two degenerate primers (P01 and P02) similar to those described by Murtha *et al.*, (1991). This primer set yielded a PCR product 120 bps in length including 75 bps of informative sequence spanning homeobox positions 61-135 (amino acid positions 21-45). For several *Hox* genes, we used the specific forward primers P03 and P06. For *Lox6*, *Scr*, *Antp*, *Lox2*, and *Lox4*, the specific primers P04 and P05 were designed to target the hexapeptide motif, a short conserved region upstream from the homeodomain which is present in all homeotic genes except *Post1* and *Post2*. Because the hexapeptide is usually separated from the homeodomain by an intron, these primers were mainly used to amplify cDNA in this study.

The PCR products corresponding to the expected sizes of the different gene fragments were excised from an agarose gel and the purified DNA fragments were cloned into a PCR 2.1-TOPO TA cloning vector (Invitrogen). The recombinant plasmids were used to transform competent *Escherichia coli* TOP10 cells and plated onto selective medium. Recombinant clones were selected by blue/white screening. DNA mini preparations of white clones were performed with a QIAprep Spin Miniprep Kit (Qiagen, Germany) according to the manufacturer's instructions. Digestion with the restriction endonuclease EcoRI was performed to ensure the presence of the inserted DNA

Purified DNA from the clones was sequenced with an ABI 310 DNA Analyzer (Perkin-Elmer) using an M13 primer. The sequences were compared using the GenBank database (NR "non-redundant") with BLASTX. Alignment of the *Hox* genes was performed using Clustal X 1.81 (Thompson *et al.*, 1997). The following GenBank sequences were for our alignment analyses (GenBank accession numbers): *Drosophila melanogaster Dme-lab* (NM_057265), *Dme-pb* (NM_057322), *Dme-Dfd* (NM_067853), *Dme-Scr* (X14475), *Dme-Antp* (NM_079525), *Dme-ftz* (NM_058159), *Dme-Ubx* (NM_080500), *Dme-abdA* (NM_057345), *Dme-abdB* (NM_080157), and *Dme-Dll* (NM_ 166689); echiuroid (*U. unicinctus*) *Uun-lab01* (AY770065), *Uun-lab02* (AY770066), *Uun-Hox3* (AY770067), *Uun-Lox6* (AY770066), *Uun-Scr* (AY770069), *Uun-Antp* (AY770070), *Uun-Lox2* (AY770071), *Uun-Lox4* (AY770072), and *Uun- Post2*(AY770073); oligochaete (Perionyx excavatus) *Pex-lab01* (AY439315), *Pex-lab02* (AY439316), *Pex-lab03* (AY439317), *Pex-pb* (AY439318), *Pex-Hox3* (AY439319), *Pex-Lox6* (AY439321), *Pex-Lox18* (AY439322), *Pex-Scr* (AY439323), *Pex-Antp* (AY439324), *Pex-Lox5* (AY439325), *Pex-Lox2* (AY439326), *Pex-Lox4* (AY439327), *Pex-Post1* (AY439328), *Pex-Post2a*

(AY439329), *Pex-Post2b* (AY439330); polychaete (*Nereis virens*) *Nvi-lab* (AF151663), *Nvi-pb* (AF151664), *Nvi-Hox3* (AF151665), *Nvi-Scr* (AF 151667), *Nvi-Dfd* (AAD46169), *Nvi-HB1* (de Rosa *et al.* 1999), *Nvi-Lox5* (AF151671), *Nvi-Lox2* (AF151669), *Nvi-Lox4* (AF151670), *Nvi-post1* (AF151672), and *Nvi-post2* (AF151673); leech (*Helobdella robusta*) *Hro-Lox6* (AF004386), *Hro-Lox5* (AF004387); leech (*Helobdella triserialis*) *Htr-Lox7* (Y10888), *Htr-Lox18* (AF137385), *Htr-Lox20* (AF006638); leech (*Hirudo medicinalis*) *Hme-Lox6* (AF017253), *Hme-Lox2* (X17566), and *Hme-Lox4* (S79240); flatworm (*Dugesia japonica*) *DjAbd-Ba* (AB0 49972) and *DjAbd-Bb* (AB049973); Cephalopoda (*Euprymna scolopes*) *Esc-lab* (AH011177), *Esc-Hox3* (AY052755), *Esc-Scr* (AY052756), *Esc-Antp* (AY052758), *Esc-Lox5* (AY052757), *Esc-Lox4* (AY052759), *Esc-Post1* (AY052760), and *Esc-Post2* (AY052761); Brachiopoda (*Lingula anatina*) *Lan-lab* (AF144672), *Lan-Hox3* (AF144673), *Lan-Scr* (AF144674), *Lan-Antp* (AF144675), *Lan-Lox5* (AF144676), *Lan-Lox2* (AF144577), *Lan-Lox4* (AF144678), *Lan-Post1* (AF144679), and *Lan-Post2* (AF144680); and bryozoan (*Bugula turrita*) *Btu-pb* (AY497421), *Btu-Hox3* (AY 497422), *Btu-DfdA* (AY497423), *Btu-DfdB* (AY497424), *Btu-Lox5* (AY497425), and *Btu-Post2* (AY497426).

REFERENCES

Austin P, F White: Metabolism and oxygen transport in the innkeeper *Urechis caupo.* Physiol Zool, 54(1): 44-54, 1981.

Cho SJ, DH Lee HJ Kwon, HA Chi, SC Park, KS Shin: *Hox* genes in the echiuroid *Urechis unicintus*. Dev Genes Evol, 216: 347-351, 2006.

Choe RS, KS Shin, CN Joo, DY Hwang: The differentiation of the female gonial cells in echiuroid *Urechis unicinctus*: A fine structural study. Korean J Electron Microscopy, 18: 77-91, 1988.

Das NK: Cytological and biochemical analysis of development of *Urechis* oocyte. Amer Zool, 16: 345-362, 1976.

De May JR: The preparation of immunoglobulin gold conjugated (IGS reagents) and their use as markers of light and electron microscopic immunocytochemisty, In: Cuello AC (ed), Immunohistochemistry, Wiley, New york, pp 347-372, 1983.

de Rosa R, JK Grenier, T Andreeva, CE Cook, A Adoutte, M Akam, SB Carroll: *Hox* genes in brachiopods and priapulids and protostome evolotion. Nature, 399: 772-776, 1999.

Elinson RP, ME Manes: Morphology of the site of sperm entry on the frog egg. Dev Biol, 63: 67-75, 1978.

Fischer A: The structure of symplasmic early oocytes and their enveloping sheath cells in the

Polychaete, *Platynereis dumerilli.* Cell Tissue Res, 160: 327-343, 1975.

Fischer A, A Dhainaut: The origin of yolk in the oocyte of *Nereis virens* (Annelida, Polychaeta. Cell Tissue Res, 240: 67-76, 1985.

Frank RR, FC Davis: Regulation of histone synthesis during *Urechis caupo* (Echiura) development. Dev Biol, 98: 101-109, 1983.

Frank RR, N Sawada, FC Davis: Histone messenger RNA synthesis and accumulation during early development in Echiuroid worm , *Urechis caupo*. Dev Biol, 109: 118-126, 1985.

Garey JR, BJ Davis: Modulation of hemoglobin from *Urechis caupo* by organic phosphates in response to low oxygen. Physiologist, 21(4): 42-46, 1978.

Garey JR, AG Riggs: Structure and function of hemoglobin form *Urechis caupo.* Archiv Biochem Biophys, 228: 320-331, 1984.

Gould-Somero M: Echiura. In "Reproduction of Marine Invertebrates". A C Giese & J S Pearse ed, 3: 277-311, Academic Press, Inc, New York, 1975.

Gould-Somero M, L Hollander: Fine structural investigation of the insemination response in *Urechis caupo*. Dev Biol, 46: 358-369, 1975a.

Gould-Somero M, L Hollander: Oocyte differentiation in *Urechis caupo*(Echiura): A fine structural study. J Morphol, 147: 475-505, 1975b.

Hall RE, RC Terwilliger, NB Terwilliger: Hemoglobins and myoglobin of the Echiuran *Urechis caupo* (Fischer and Macginite). Comp Biochem Physiol, 708: 353-357, 1981.

Hoffmann, NH, CP Magnum: The function of the coelomic cell hemoglobin in the polychaete, *Glycera dibranchiate*. Comp Biochem Physiol, 36: 211-228, 1970.

Horowitz, NH, JP Baumberger: Studies on the respiratory pigment of Urehis egg. Biol Chem, 141: 407-415, 1941.

Jeong JW, RS Choe, KS Shin, CN Joo: The differentiation of the male gonial cells in the storage sac in Echiuroid, *Urechis unicintus*: A fine structural study. Korean J Electron Microscopy, 19: 70-88, 1989.

Korschelt E, K Heider: Lehrbuch der Vergleichenden Entwicklungsgeschlechte der Tiere. Jena, Gustav Fischer, 1902.

Kwon HJ, WJ Kim, KS Shin: Fine structural investigations of fertilization envelopes and acrosomal reaction in *Urechis unicinctus*. Korean J Electron Microscopy, 30(1): 6-72, 2000.

Kwon HJ, JW Jeong, WJ Kim, KS Shin: Immunofluorescence and electron microscopic study on the artificial insemination and rotation-shift behaviors of the bipolar spindle fiber in *U. unicinctus* egg. Korean J Electron Microscopy, 33(2):105-116, 2003.

Mangum P, RC Terwilliger, NB Terwilliger, R Hall: Oxygen binding of intact cells and extracted hemoglobin of the Echiuran *Urechis caupo*. Comp Biochem Physiol, 76A(2): 253-257, 1983.

Masui T: D-galactoside specific lectins from coelomogyte of the Echiuran, *Urechis unicinctus*. Biol Bull, 166: 178-188, 1984.

McHugh D: Molecular evidence that echiurans and pogonophorans are derived annelid. Proc Natl Acad Sci USA, 94:8006-8009,1997.

Miller JH: Studies of oogenesis in *Urechis caupo:* Part I. Methods for separating different size classes of oocytes. Dev Biol, 32(1): 219-223, 1973.

Muneoka Y, M Kamura: Actions of noradrenalin and some other biogenic amines on the body wall muscles of an echiuorid *Urechis unicinctus*. Comp Biochem Physiol Com Pharmacol, 72(2):281-288,1982

Murtha MT, JF Lechman, FH Ruddle: Detection of homeobox genes in development and evolution. Proc Natl Acad Sci USA, 88: 10711-10715, 1991.

Newby WW: The embryology of the echiuroid worm *Urechis caupo*. Men Am Phil Soc, 16: 1-19, 1940.

Ochi, O: Blood pigments and erythrocytes found in some marine annelida. Mem Am Ehime, Univ Sci Ser B, 6: 23-91, 1969.

Paul M: The polyspermy block in eggs of *Urechis caupo*: evidence for a rapid block. *Exp Cell Res,* 90: 137-142, 1975.
Pritchard A, FN White: Metabolism and oxygen transport in the inkeeper *Urechis caupo*. Physiol Zool, 54(1): 44-54, 1981.
Remana A, V Storch, U Weisch: Systematische Zoologie. Staemme Tierreichs. 28pp. Gustav Fischer Verl. Stuttgart, 1975.
Richardson KC, L Jarett , EH Finke: Embedding in epoxy resins for ultrathin sectioning in electron microscopy. Stain Technol, 35: 313-323, 1960.
Rouse GW, K Fauchald: Cladistics and polychaetes. Zool Sci, 26: 139-204, 1997.
Sato H: Studies on the echiuroidea, sipunculoidea and priapuloidea of Japan. Sci Rep Tohoku Imp Univ Ser IV, Vol XIV, pp 339-460,1939.
Sawada N, K Ochi: Electron microscope studies on sperm differentiation on marine annelid worms. Part I: Sperm formation in *Ikedosoma gogoshimens*. Developm Growth Differ, 17(2): 77-88, 1975.
Shin KS, HJ Kwon: The differentiation of coelomocytes according to the cyclic reproduction periods of *Urechis unicinctus*. Soonchunhyang J Nat Sci, 3(2): 623-634, 1997.
Shin KS, SH Lee: Ultrastructure and filtrating function of the ciliated epithelial cells of foregut in *Urechis unicinctus*. Korean J Electron Microscopy, 28: 435-446, 1998.
Shin KS, HJ Kim, WJ Kim: Localization of anti-actin-gold particles (10nm) labeled to nuclear actin of *Urechis* sperm and spermatid. Korean J Electron Microscopy, 30(4): 403-412, 2000.
Shin KS, HJ Kim, HJ Kwon, WJ Kim: Cytochemical localization of nuclear actin of sperm and spermatids in *Urechis unicinctus*. Integrative Bioscience, 9: 65-73, 2005a.
Shin KS, HJ Kwon, WJ Kim: Attribution of cortical granules to formation of fertilization envelope and polyspermy block in *Urechis unicinctus.* Integrative Bioscience, 9: 57-64, 2005b.
Shin KS, WJ Kim: Fine structural study of coelomic solitary spermatogenesis in *Urechis unicinctus*. Korean J Electron Microscopy, 32(2): 107-119, 2002.
Slot JW, HJ Geuze: Immunolabelling for electron microscopy. In "Techniques in immunocytochemistry" vol. 4 G. R. Bullock ed, pp 67-93, Elsevier Science, 1989.
Terwilliger R, R Garlick: Hemogobin of *Glacerarobusta*: Oxygen equilibrium properties of coelomic cell hemoglobin and body wall myoglobin. Comp Biochem Physiol, 59A: 359-362, 1978.
Thompson JD, TJ Gibson, F Plewniak, F Jeanmougin, DG Higgins: The ClustalX windows interface: flexible strategies for multiple sequence alignment aided by quality analysis tools. Nucleic Acids Res, 25(24): 4876-4882, 1997.
Yamamoto M, M Okawa, J Ishida: An electron microscopic study of oogenesis in the Echiuroid, *Urechis unicinctus*. J Fac Sci Tokyo Univ. Sec IV, 13(1): 39-54, 1973.

Chapter 02

Urechis coelomocytes

Chapter 02

Urechis coelomocytes

2-1 Introduction

It is known that in *U. unicinctus,* some internal organs and cells ordinarily reside in the coelomic cavity whereas the appearance of others changes according to seasons. The latters include storage sacs, nephridia, and sex cells that appear only during the breeding season. The formers include digestive tract that contains the respiratory organ (hind gut), and coelomocytes floating in body fluid. Macrophages are not regarded as ordinary residents because they appear only when the worm is infected (Miller, 1973; Garey & Riggs, 1984; Masui, 1984; Frank & Davis, 1983; Frank *et al.*, 1985).

The seasonal appearance of certain organs and cells is apparently related to the reproductive cycle. The coelomocyte is an unique cell sharing the coelomic cavity and fluid, in which the gametes are developed in the breeding season, it is necessary to study the function and role of the coelomocyte to understand *Urechis* reproductive cycle and gametogenesis.

Diffuse coelomocytes are known as blood cells responsible for synthesizing hemoglobin molecules (Hb) and respiratory metabolism (Fisher *et al.*, 1928; Ochi, 1969: Hoffmann & Magnum, 1970; Terwilliger & Garlick, 1978; Pritchard & White, 1981). In the breeding season, these cells are believed to play some key roles in the reproductive cycle because they are enclosed in the same tissues and organs with developing sex cells of thc worms. As is described above, the coelomocytes are unique cells residing in the coelomic cavity along with *Urechis* sex cells (Choe *et al.*, 1988; Kwon, *et al.,* 2000; Jeong *et al.*, 1989).

Previous studies revealed that coelomocytes do not stay still inactively in every alternation of the seasons holding the coelomic cavity in common with developing reproductive cells. The number and structural differentiation of coelomocyte vary greatly with the reproductive cycle. The number of coelomocytes is greatly increased in the breeding seasons, and returns to normal levels out of the season. The surface of the coelomocytes is decisively differentiated in structure during the breeding seasons, and completely returns to normal out of the season.

It has been speculated that the contents of coelomocyte differ in form and substance. Coelomocytes are one of the coelomic constituents that fluctuate and vary in cell number and morphology according to seasonal changes similar to other components. Seasonal changes of the coelomocytes are similar to those of somatic cells during the formation of tissues or organs together with sex cells in other animal species.

Coelomocytes do not appear to be limited to a single function, but are multifunctional cells involved in normal life cycles as well as reproduction. Previous investigations studying life and reproductive cycles revealed that seasonal changes of the coelomocyte are related with functional amplification or adaptive changes involved in growth and reproduction of the worm.

In this section, the structural and functional transformation of coelomocytes is described, and the multifunctional role of these cells related to coelomocyte nomenclature is outlined. The term "coelomocyte" does not seem to reflect the functions and roles of this cell in the cyclic life, but was derived based merely on its phylogenic senses.

2-2 Seasonal fluctuations of coelomocyte number and structure

Except for the digestive tract, the coelomic cavity is filled with coelomocytes and coelomic fluid out of the breeding season. The coelomocytes are single, free-floating, and unique cells contained in body fluid, and account for half the ratio of the body fluid in a gravity field (300 × g). In the presence of sex cells, the number and structure of coelomocytes dramatically changes. Out of the breeding season (September to December), the number of cells is 60,000/mm^3; but during the peak breeding season, this number increases up to 167%, reaching maximum number of 100,000/mm^3 (mid-April to May).

Sex cells begin to appear during mid-January and increase in number at an exponential rate; this is proportional with increases in the number of coelomocytes. Decreases of both sex cells and coelomocytes also occur co-incidently at the end of the breeding season. However, the sex cells dis-appear completely from the coelomic cavity after the spawning, while the number of coelomocyte decreases promptly until they reach normal levels.

2-3 Structural variations of coelomocytes

Increased numbers of coelomocytes are accompanied by structural changes in these cells. The changes that occur during the breeding seasons are very dramatic. In fact, the coelomocytes can actually appear to be an-other cell type at first glance during this time.

Coelomocytes can be classified into three groups based on season-al-related differentiation. Group A is coelomocytes prepared out of the breeding season (Fig. 2-1, 2-2). Group B cells are formed during the breed-ing season and are further subdivided into groups B_1 and B_2 according to the preparation timing. B_1 cells are prepared during the first half of the breeding season (from January to the beginning of March) (Fig. 2-3, 2-4). B_2 cells are produced during the latter half of the season including the high-er breeding season (from mid-April to the spawning season) (Fig. 2-5, 2-6). Altogether, structural differentiation of the coelomocytes seems closely as-sociated with growth and development of the sex cells.

2-3-1. Group A coelomocytes

Coelomocytes are nucleated cells and are not enclosed in compact tissues. The cells often form aggregates containing several cells, but it seemed that these aggregations are not involved with tissue formation. In preparative experiments, coelomocyte aggregations are easily dissociated into individual cells. The tendency of the aggregation is not known, but cell aggregations can increase in number and size, especially under laboratory conditions.

The coelomocytes are slightly polygonal in shape with 8 μm diameter (Fig. 2-1, 2-2). The nuclei are ellipsoid and relatively large with a nucleo-cytoplasmic ratio of approximately 1:1. They contain heterogeneous nuclear matrices. The cell organelles are not well developed, but mitochondria and rER are discernible in the endoplasm. Free ribosomes are scattered throughout the exoplasm. The cell surfaces are rather smooth and simple with a number of small protrusions.

2-3-2. Group B_1 coelomocytes

Group B_1 cells are thought to be the intermediate form of A type cells transforming into B_2 type cells (Fig. 2-3). During the breeding season, the sex cells appear and increase in number. The sex cells are resembled morphology in general and size to coelomocyte, confused often with the coelomocyte, except the transparency and tendency of aggregation one another.

The sex cells of both genders appear in coelomic fluid as aggregates con-

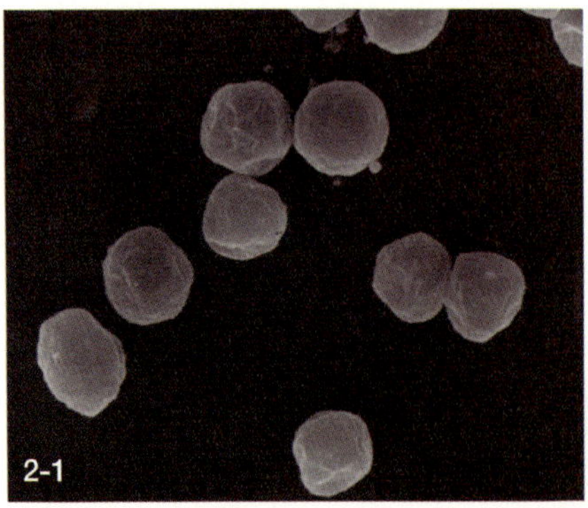

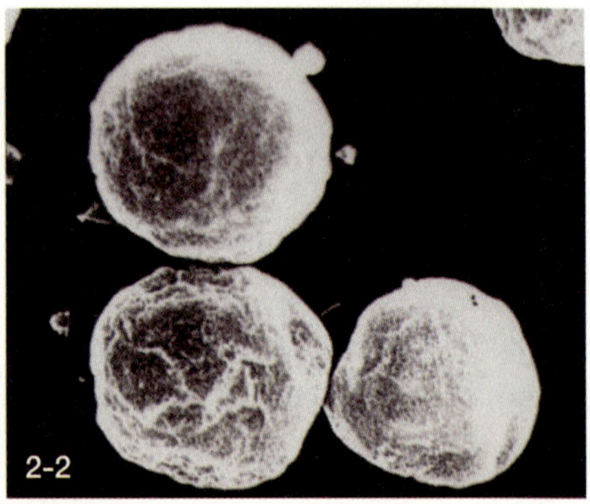

Fig. 2-1 Out of breeding season, SEM observation reveals that coelomocytes are smooth surfaced with a few exocytotic vesicles and secretions. × 1,100.

Fig. 2-2 Negative staining of coelomocytes prepared out of the breeding season exhibit slightly rectangle shape of the coelomocyte. A few exocytotic vesicles are to be seen, and cell surface is not differentiated. × 20,000.

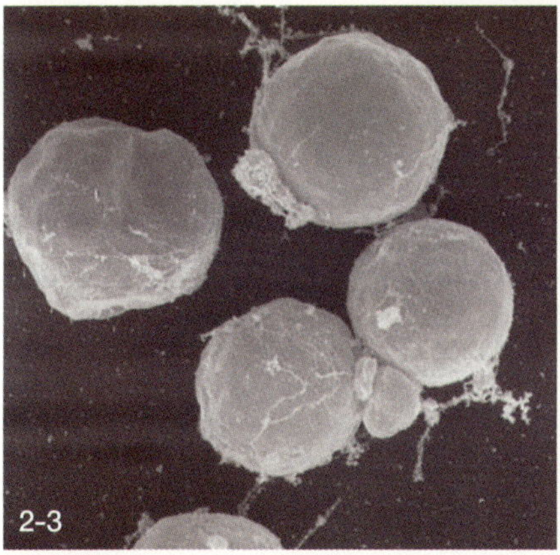

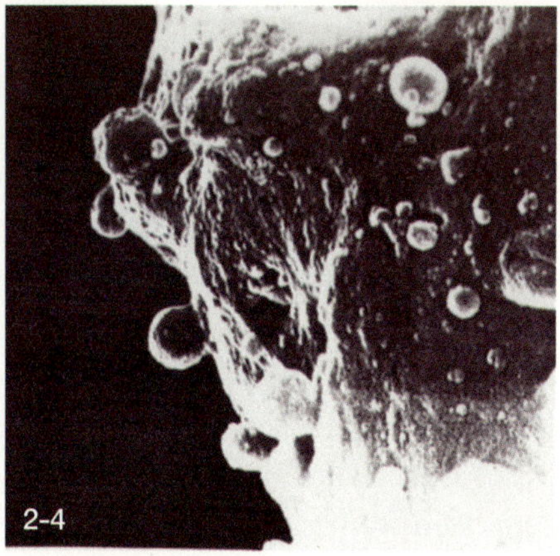

Fig. 2-3 Approaching close to breeding season, differentiations of exocytotic vesicles and appearances of fibrous materials are observed on the coelommocyte surface. × 20,000.

Fig. 2-4 Exocytotic vesicles on the coelomocyte surface are increased in number parallel with the growth and differentiation of the gametes sharing the coelomic cavity with the coelomocytes. × 50,000.

taining several cells. At appearance, the sex cells are smaller (6~7 μm), transparent, and tend to assemble into cellular cluster while coelomocyte are larger (8 μm), opaque, and dispersed in the coelomic fluid. Although the sex cells separate from each other over time, transparency along with additional unique characteristics of the sex cells that distinguish them from other cells are maintained.

SEM examination showed that a number of protrusions or exocytotic vesicles of various shapes and sizes arise from the surface of the cell (Fig. 2-4). A network of fine fibrous materials covers the entire surface of the cell and vesicles. Approaching the higher season, the vesicles are increased in number and the fibrous materials covering the cell surface are raised in amount. Using EM, many microspheres or microbodies of various densities and size less than 1 μm in diameter were found in the cytoplasm. The density of the nuclear matrices also increases significantly.

2-3-3. Group B_2 coelomocytes

The higher breeding season of the worm is defined according to the sex cell development, as the time periods at which some mature sex cells are observed in the coelomic fluid while some are stored in the storage sac. During the higher breeding season, changes of coelomocytes are observed mainly on the surface of group B_2 cells. Numerous exocytotic vesicles of various shapes (1-3 μm in diameter) protrude from the B_2 cell surface. Abundant fine fibrous materials cover the entire surface of the cells. Serial observations have shown that B_2 cells are derived from B_1 cells, and ultimately originate from A group cells.

It is difficult to separate B_2 cells from the sex cells in gravity fields, partly due to the physical properties of the cells and by entanglements of the fibrous materials (Fig. 2-5, 2-6). Coelomocytes of the A groups are easily separated from sex cells in the same gravity field of 300 × g. The bottom coelomocyte is sharply separated from the top sex cell layer. Under these same conditions, B_1 coelomocytes are not difficult to separate from the sex cell, even though the border line is not clear as the case of the A groups cells.

Coelomic fluid composition has been analyzed to understand mechanisms of the coelomocyte differentiation especially in parallel with sex cell development. Coelomic fluids undergo changes mainly in amino acid composition. The levels of amino acids such as Asp, Thr, Ser, Val, Met, Ile,

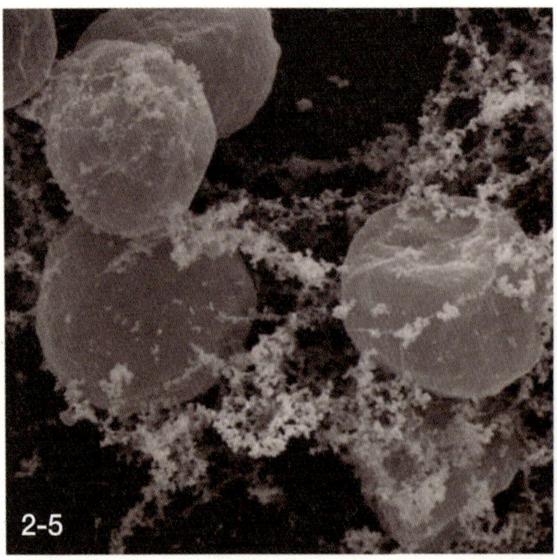

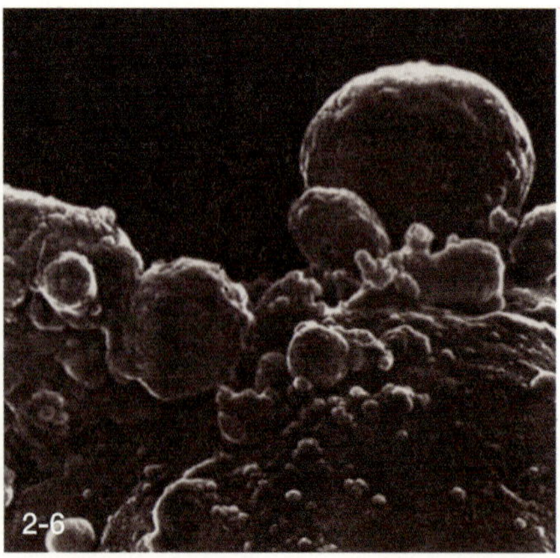

Fig. 2-5 In the higher breeding season, the entire surfaces of coemocytes are covered with entangled fibrous materials derived actually from the coelomocyte secretions. × 25,000.

Fig. 2-6 Secretions of coelomocytes seem heterogeneous including the fibrous, humoral and vesicular secretions. Number of coelomocyte increases up to 167% during the peak breeding season. × 12,000.

Leu, Lys and Tyr along with ammonia and beta-amino-isobutyrate concentration increase considerably in amount (Fig. 2-7, 2-8). On the other hand, there is EM evidence that part of the exocytotic vesicles detached from B_2 cells being released into body fluid. The compositional changes of coelomic fluid seem closely related to the secretory activities of B_2 cells. At the beginning of the higher season, sex cells enter a phase of the exponential growth. For instance, oocytes increase in volume 5,000-fold mainly due to the deposition of yolk. At the same time, a considerable number of sperm are flagellated and undergo spermiogenesis. The cells of the body and intestinal wall are not exclusive of the materials required for sex cells growth.

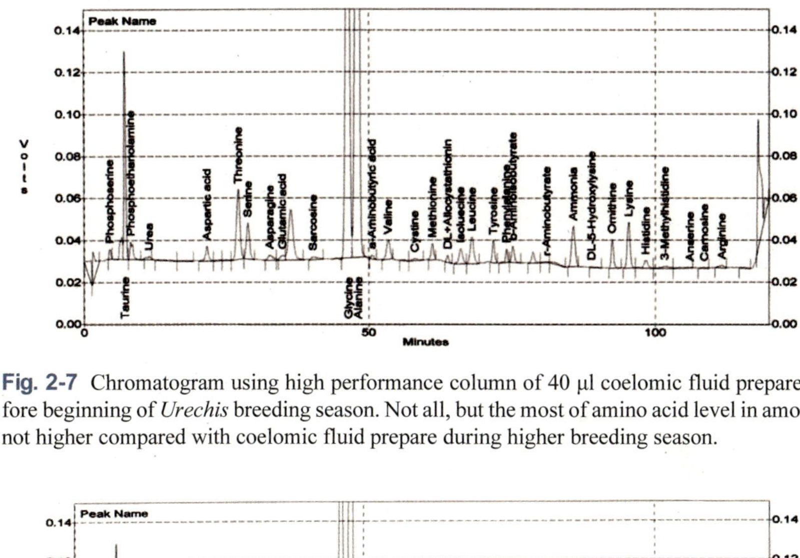

Fig. 2-7 Chromatogram using high performance column of 40 μl coelomic fluid prepared before beginning of *Urechis* breeding season. Not all, but the most of amino acid level in amount is not higher compared with coelomic fluid prepare during higher breeding season.

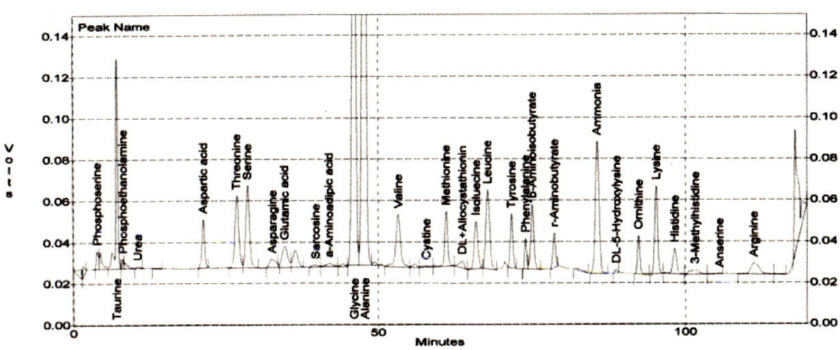

Fig. 2-8 Some amino acid components, Asp, Thr, Ser, Val Met, Ile, Leu, Lys and Tyr increase in amount in continuation of the breeding season. Compositional change is corresponded to structural differentiation producing secretory vesicles of coelomocyte surface.

However, B_2 cell secretions seem to provide a certain substantial basis for exponential sex cell growth. The differentiation and increased number of B_2 cells are regular events that occur during every reproductive cycle of the worm. B_2 coelomocytes can be defined by their role in the growth and differentiation of sex cells.

2-4 Function of coelomocyte

The nomenclature of a cell is normally based on cellular functions. However, the term "coelomocyte" does not reflect the function or origin of this type of cell. It merely indicates in traditional and phylogenic sense that this is a free single cell that normally resides in the coelomic cavity (Terwillger & Garlick 1978). In follow-up studies, the coelomocyte was referred to as a "blood cell" (Fisher & MacGnite, 1928; Ochi, 1969; Horowitz & Baumberger, 1941; Hoffmann & Magnum, 1970) responsible for synthesizing and retaining hemoglobin molecules (Hb) and respiratory metabolism (Ochi, 1969: Hall *et al*., 1981; Magnum *et al.,* 1983; Garey & Davis, 1978). On the other hand, after extracting Hb molecules from coelomic fluid, others groups seemed to not approve this term. It is reported that Hb molecules are synthesized in coelomocytes and discharged into the coelomic fluid where they are directly involved in respiratory metabolism. Thus, coelomic fluid may play a practical role similar to that of blood (Hoffmann & Magnum, 1970; Hall *et al*., 1981; Magnum, 1983; Garey & Riggs, 1984). Another secretory function is closely related to sex cell development. Some coelomocyte secretions enter the developing oocytes. Subsequently, the oocyte volume is greatly increased due to the accumulation of yellow yolk in the cytoplasm (Shin *et al*., 2004). Hb molecules have been isolated from extracts of developing oocytes. The Hb molecules seem entered into the growing oocyte and the absorbed substance is known as "urechrome" since it is a source of pigmentation of the coelomic fluid (Hoffmann & Magnum, 1970).

The term *Urechis* "blood" now refers to coelomic fluid containing Hb molecules. The term "blood cell" is used for coelomocytes that synthesize the Hb molecules. This contradiction of terms seems to be due to differences in the synthesis and operating site of the Hb molecules. As previously mentioned, this work uses the term "coelomocyte" and "coelomic fluid" rather than "blood cell" and "blood" when referring to coelomic fluid containing Hb molecules and its cellular components. This is because coelomocytes do not apparently regulate the function of secretions containing Hb, but are involved in particular purposes of the sex cell growth and developments. The coelomocytes are not set limit to the function of the Hb secretions, but involved in particular purposes of the sex cell growth and developments. "Urechrome" is the prevailing term used to describe the different coelomocyte functions, and Hb molecules appear to be included in various coleomocyte secretions.

One of the important consequences of oogenesis is the production of cellular components that will be utilized during post-fertilization development. However, most oocyte genes are inactivated during oogenesis except for ones involved in RNA synthesis (Das, 1976; Davidson *et al*., 1982). Since the yolk is composed of common phospholipids (Fischer, 1975; Chelmon & Sizou, 1986), the yolk or yolk precursors are probably derived ultimately from sources outside of the oocytes, whether the yolk is produced by autosynthesis (Anderson, 1964; Anderson, 1969), or heterosynthesis (Baert & Solomianny,1987; Ekelbarger, 1979; Fischer & Dhainaut, 1985). *Urechis* oocytes do not appear to be exceptional. Initially, the oogonium is 6~7 μm in diameter. At the time of spawning, the diameter of the mature oocyte is 150 μm (Kwon *et al*., 2000). The oocyte volume increases at least 15,000 times during oogenesis, primarily due to yolk deposition. It is likely *Urechis* yolk is derived from coelomocyte secretions. Coelomocytes are transformed into secretory cells and secretion is not limited to the release of Hb molecules. Substantial changes of amino acid composition in coelomic fluid are associated with oocyte growth and development. Coelomocyte structure also undergoes changes including the appearance of numerous protuberances or exocytotic vesicles on the surface of the B_2 cells. These changes may be an evidence of transformation into secretory active cells. It seems logical to speculate that coelomocytes increase 167% in number and undergo transformation into secretory cells to support the development of the sex cells.

It is common in gametogenesis, changes in the cell number and structure that parallel sex cell development can be seen in accessory cells that support the sex cells developments (Yasumasu *et al*., 1975; Tazawa & Yasumasu, 1977; Palombi *et al*., 1979). In *Urechis*, transformation of the coelomocytes offers the causal relationship of the accessory cells to the differentiating sex cells. Increases in the number of coelomocytes are believed to represent functional amplification of individual coelomocytes while structural changes are considered to reflect modified functions of the coelomocytes. The functional amplification can be defined as the final consequence of sex cell development, but the modified functions are involved in the production of secretions that support the sex cells.

Summary

Coelomocyte is unique cell in *Urechis* coelomic cavity out of breeding season. During breeding season, the coelomocyte shares the coelomic cavity with developing male and female gametes. Including the spatial relationship with gametes, the coelomocyte fluctuates in number and structural differentiation. During the higher breeding season, number of coelomocyte increases up to 167% and numerous exocytotic vesicles of various sizes appear on the surface of the coelomocyte, transforms to secretory active cell.

Coelomocytes have multiple functions that change according to seasonal alternations. Out of the breeding season, coelomocytes are responsible for synthesizing and secreting Hb molecules. During the higher breeding season, transformed coelomocytes seem to serve as accessory cells that support and promote gamete developments. It can therefore be concluded that coelomocytes are a type of secretory cell. Its proper function is the synthesis and secretion of Hb molecules while its auxiliary function is related to accessory cell in support for sex cell development. The name of such kind of cell could therefore not be based on a particular selective or specialized function. Thus, phylogenic properties are more suitable for defining the nature of coelomocytes, seem to be more inclusive of functional variations that occur according to seasonal changes.

REFERENCES

Anderson E: Oocyte differentiation and vitellogenesis in Roach *Periplaneta Americana*. J Cell Biol, 20: 131-155, 1964.

Anderson LM, WH Telfer: The follicle cell contribution to the yolk spheres of Moth oocytes. Tissue Cell, 1: 633-644, 1969.

Baert JL, MC Slomianny: Heterosynthetic origin of the major yolk protein vetellin in a Nereid *Prenereis cultirfera*(Polychaete annelid). Comp Biochem Physiol, 88B: 1191-1199, 1987.

Choe RS, KS Shin, CN Ju, DY Hwang: The differentiation of the female gonial cells in Echiuroid *Urechis unicinctus*.: A fine structural study. Korean J Elect Micros, 18: 77-91, 1988.

Chelmon VP, AV Sizov: Lipid matrix of the erythocyte membrane of marine worm *Urechis nicinctus* (Von Drashe). Comp Biochem Phsiol, 83B: 241-244, 1986.

Davidson EH, BR Hough, RJ Britten: Molecular biology of the sea urchin embryo. Science, 217: 17-26, 1982.

Das NK: Cytological and biochemical analysis of development of *Urechis* oocyte. Ann Zool, 16: 345-362, 1976.

Ekelbarger KJ: Ultrastuctural evidence for both autosynthetic yolk formation in the oocytes of an annelid(*Phragmatopoma lapidosa*: Polychaeta). Tissue Cell, 11: 425-443, 1979.

Fisher, WK, GB MacGinite: The natural history of Echiuroid worm. Ann Mag Nat Hist, 10: 204-213, 1928.

Fischer A: The structure of symplasmic early oocytes and their enveloping sheath cells in the Polychaete, *Platynereis dumerilli.* Cell Tissue Res, 160: 273-343, 1975.

Fischer A, A Dhainaut: The origin of yolk in the oocyte of *Nereis virens*(Annelida , Polychaeta. Cell Tissue Res, 240: 67-76, 1985.

Frank RR, N Sawada, FC Davis: Histone messenger RNA synthesis and accumulation during early development inEchiuroid worm , *Urechis caupo*. Dev Biol, 109: 118-126, 1985.

Frank RR, FC Davis: Regulation of Histone synthesis during *Urechis caupo* (Echiura) development. Dev Biol, 98: 101-109, 1983.

Garey JR, BJ Davis: Modulation of hemoglobin from *Urechis caupo* by organic phosphates in response to low oxygen. Phsiologist, 21(4): 42-46, 1978.

Garey JR, AG Riggs: Structure and function of hemoglobin form *Urechis caupo* Archiv. Biochem Biophys, 228: 320-331, 1984.

Hall RE, RC Terwilliger, NB Terwilliger: Hemoglobins and myoglobin of the Echiuran *Urechis caupo* (Fischer and Macginite). Comp Biochem Physiol, 708: 353-357, 1981.

Horowitz, NH, JP Baumberger: Studies on the respiratory pigment of *Urehis* egg. Biol Chem, 141: 407-415, 1941.

Hoffmann, NH, CP Magnum: The function of the coelomic cell hemoglobin in the polychaete, *Glycera dibranchiate*. Comp Biochem Physiol, 36: 211-228, 1970.

Jeong JW, RS Choe, KS Shin, CN Joo: The differentiation of the male gonial cells in the storage sac in Echiuroid, *Urechis unicintus*: A fine structural study. Korean J Electron Microscopy, 19: 70-88, 1989.

Kwon HJ, WJ Kim, KS Shin: Fine structural investigations of fertilization envelopes and acrosomal reaction in *Urechis unicinctus*. Korean Electron Microscopy, 30(1): 6-72, 2000.

Mangum P, RC Terwilliger, NB Terwilliger, R Hall: Oxygen binding of intact cells and extracted hemoglobin of the Echiuran *Urechis caupo*. Comp Biochem Physiol, 76A(2): 253-257, 1983.

Masui T: D-galactoside specific lectins from coelomogyte of the Echiuran, *Urechis unicinctus*. Biol Bull, 166: 178-188, 1984.

Miller JH: Studies of oogenesis in *Urechis caupo*: Part I. Methods for separating different size classes of oocytes. Dev Biol, 32(1): 219-223, 1973.

Ochi, O: Blood pigments and erythrocytes found in some marine annelida. Mem Am Ehime, Univ Sci Ser B, 6: 23-91, 1969.

Pritchard A, FN White: Metaboism and oxygen transport in the inkeeper *Urechis caupo*. Physiol Zool, 54(1): 44-54, 1981.

Palombi F, E Ziparo, FF Rommerts, JA Grootegoed, M Antonini, M Stefanini: Morphological characteristics of male germ cells of rats in contact with Sertoli cells *in vitro*. J Reprod Fert, 57: 325-330, 1979.

Shin KS, DH Lee, TY Ko: Cleavage pattern of *Urechis unicinctus* eggs in vitro fertilization. Korean J Electron Microscopy, 34(1): 71-81, 2004.

Terwillger R, R Garlick: Haemogobin of *Glacerarobusta*: Oxygen equilibrium properties of coelomic cell haemoglobin and body wall myoglobin. Comp Biochem Physiol, 59A: 359-362, 1978.

Tazawa E, I Yasumasu: Enzyme activities in the glycolysis system of Urehis eggs with special reference to the activation of its rate-limiting enzymes following fertilization. Exptl Cell Res, 108: 191-199, 1977.

Yasumasu I, E Tazawa, A Fujiwara: Glycolysis in the eggs of Echiuroid, *Urechis unicinctus* and the oyster, *Crassaostrea gigas*. Exptl Cell Res, 93: 166-174, 1975.

Chapter 03

Ciliary filtration

Chapter 03 Ciliary filtration

3-1 Ciliary filtration

3-1-1. Introduction

Many tidal and aquatic animals filter food material and prey by external or internal ciliary movement. Animals with a filter apparatus are collectively called filter feeders without considering taxonomical status (Lynn, 1991). Filter feeders include some of echinoderms, mollusks, sponges, and annelids (*Ciliophora:* Hannes & Foissner, 1992), *Tetrahymena* (Akowska, *et al.,* 1982), *Sorogena* (Bradbury & Olive, 1980) and *Ciliates* (Hofmann-Muenz, 1991).

Member of the genus *Urechis* can be classified as annelids, echinoderms or mollusk, but this classification is not precisely fixed. These worms are coelomates that inhabit tideland, and have a well-organized anatomy with a body wall and coelomic fluid containing myosin and hemoglobin molecules, respectively. Previous ecological observations indicated that *Urechis* are filter feeders (Edmond & Stephan, 1972; Rolf, 1985). An external filter is not visible on the proboscis, leading to the suggestion that the filter apparatus of the worm is an internal organ (Rolf, 1985).

In filter feeders, it is common for the digestive tract is divided into the intestine and water duct though which filtered water flows out of the organism. The digestive tract of *Urechis* is linear and does not include a separate water duct. The linear digestive system of *Urechis* can be divided into three sections according to function: the foregut (esophagus), midgut, and hindgut. The foregut functions as a filter, the midgut is the actual digestive organ, and the hindgut serves as a respiratory organ (Remana, *et al.*, 1975; Austin & White, 1981).

Numerous ligaments connect the body wall to the surface of the digestive tract including the esophagus. Negative pressure (Mangum *et al.*, 1983) created by elaborated muscular movement is transmitted to the esophagus and causes the esophageal filter apparatus to contract and expand. During esophageal expansion, a large quantity of sea water flows into the esophagus. The filter apparatus is composed of ciliated and other kinds of cells. Numerous cilia covering over the inside surface of the esophagus move food materials contained in the absorbed sea water to the bottom of the esophageal orifice. The cilia are highly differentiated kinocilia equipped with a kinetosome, axosome, basal plate, accessory root, and main root. The main root of the kinocilia is elongated and extends into the endoplasm of the cilia cell.

The esophagus is composed of three types of cells: ciliary, mucous and other kind of secretory cells. The tissue of the esophagus is organized into pseudostratified ciliated columnar epithelium. However, ciliary beating alone is not enough to filter food materials. Negative and positive pressures created by muscular movement are also required filtering. The esophagus has been closely observed to understand the food-filtration mechanism from the inflows of sea water (Shin & Lee, 1998).

3-1-2. Tissue organization of the filtration apparatus

At creation of positive pressure of body wall, the esophageal lumen appears narrow with repetitious convex and concave patterns in the pseudostratified ciliated cell layer with numerous cilia projecting into the lumen (Fig. 3-1, 3-3). When negative pressure is created by body wall, the inside of foregut (esophageal) is widened and the lumen is opened with retracted ciliary layer (Fig. 3-2, 3-4).

The cilia are 16~18 μm in length, and form a network in the lumen by entangling when the body muscle contracts and sea water flows out of the esophagus. Food particles and debris in number are trapped in this network of cilia. Three kinds of cells compose the esophageal tissue: ciliary (16 × 50 μm), mucous and other types of secretory cells containing lipid droplets

(Fig. 3-2). In reflection of electron optics, it looks like as saturated lipid. All kinds of cells form the mucous layer of the epithelial tissue underlying submucousal layer.

Mucous cells are polarized and contain many granules (1 μm in diameter) mainly in the apical portion of the cell (Fig. 3-5). The nucleus and organ-

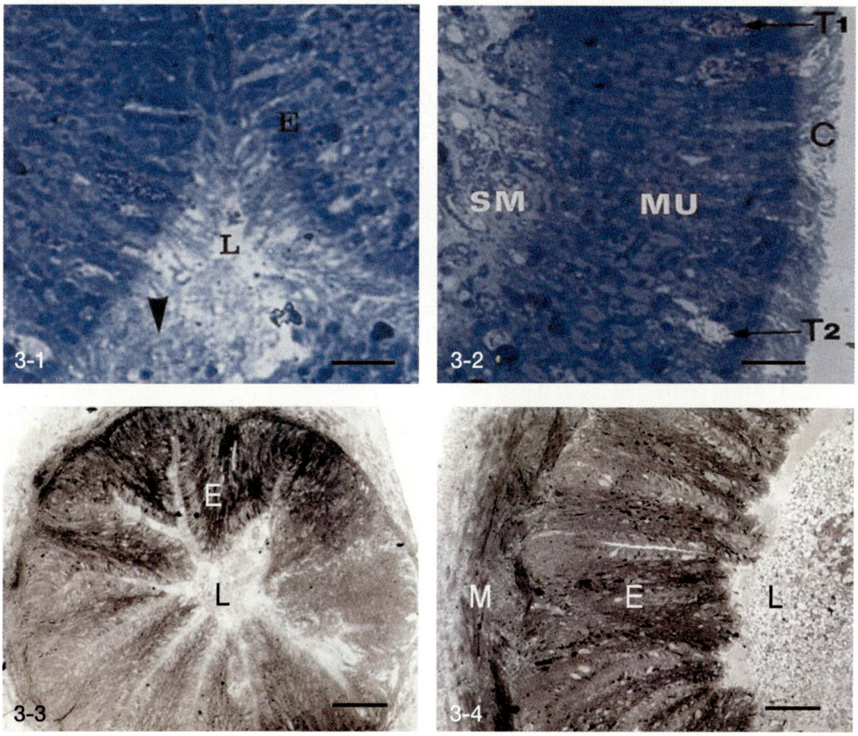

Fig. 3-1 Cross section at level of the esophagus reveals a narrowed lumen (L) by the constricted ciliated epithelial cell layers (E). The numerous cilia are densely packed in the narrowed lumen. The packed cilia and ciliary movements separate out selectively planktonic foods (arrow head) and sand particles. Bar: 10 μm.

Fig. 3-2 Foregut is composed mainly of ciliated cells and secretory cells (T1, T2) embedded in the ciliated columnar epithelial cell cells. Cell layer is consisted with the mucous (Mu) and submucous ayer (SM). C: cilia, Bar: 10 μm.

Fig. 3-3 When positive pressure is applied on foregut, the lumen (L) is narrowed with numerous cilia and ciliated epithelial cells (E) of the esophageal wall. In the lumen, some amorphous materials including foods are trapped in entangled cilia. Bar: 15 μm.

Fig. 3-4 Negative pressures created by body muscle expand the esophageal wall (E) causing esophageal lumen (L) open. Sea water containing foods flows into the expanded lumen. Bar: 15 μm.

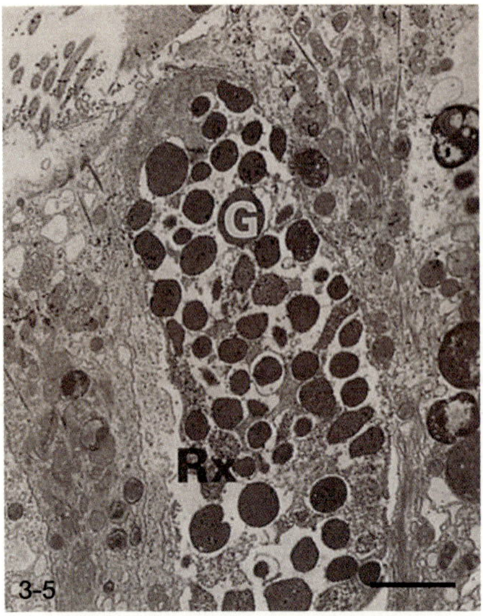

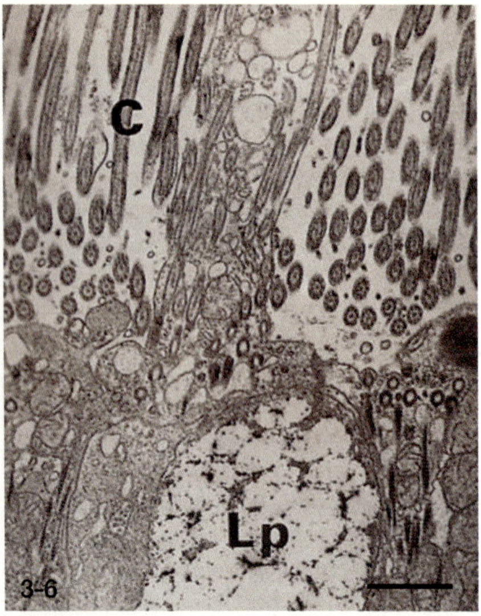

Fig. 3-5 The Type 1 secretory cell embedded in the epithelial cell layer is goblet cell containing mucous granules (G) and granular matrix (Rx) in the apical portion of the cell. Bar: 2 μm.
Fig. 3-6 The Type 2 cell is embedded in the ciliary cells (C), appeared as a cell secreting lipid substances (Lp) on the cell layer. Bar: 2 μm.

elles are located at basal portion of the cell. The granules are similar to ones in the goblet cells of vertebrates and participate in mucus secretion.

Cilia cells are also polarized with the nucleus (4 × 8 μm) located in the base of the cell. The organelles are in the apical portion of the cell (Fig. 3-6). The cell surface is covered with about 200 cilia per cell. Lipid cells are found frequently embedded between cilia cell. EM studies have concentrated on ciliary cells to characterize the nature and structure of the cilia. In fact, the number of secretory cells is small in number compared to the ciliary cells.

Esophageal cilia contain three parts: the cilia, kinetosome, and ciliary roots (Fig. 3-10). The kinetosome (220-900 nm in length) is located on the surface of ciliary cells between the cilia and ciliary root. The ciliary microtubules are arranged in 9+2 pattern. Doublet and singlet microtubules are fused together to form axosome at the outer boundary of the kinetosome. The axosome is continuous with a kinetosomal fiber surrounding the interior of the kinetosomal hole (170 nm in diameter). The kinetosomal fibers are fused together once more at the basal plate inside boundary of the kinetosome from which main and accessory roots of the cilia to extend into the cytoplasm (Fig. 11). In a practical sense, the kinetosome is the portion between the axosome and basal plate (Rudman, 1972) and is cylinder a surrounded by kinetosomal fibers. The cell membrane becomes invaginated as it approaches the kinetosome and covers two-thirds of the kinetosome. The cell membrane is continuous and covers the entire cilia. In the cross-section

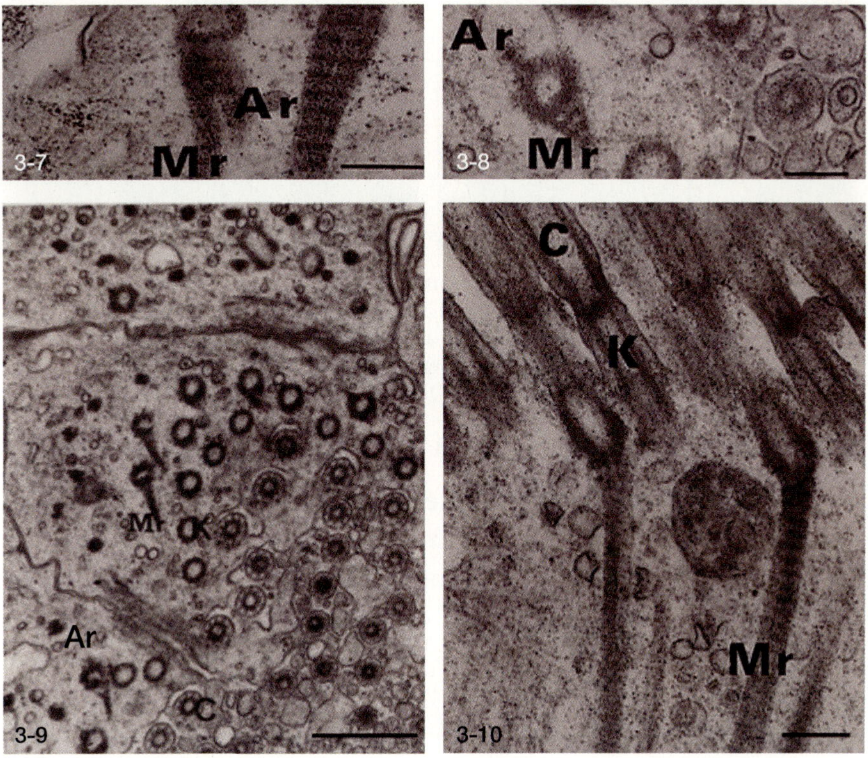

Fig. 3-7 An oblique section through the level of kinetosome and plasma membrane of the ciliary cells reveals the beginning portion of the cilia roots. The ciliary roots are directly connected with basal plate of the kinetosome, and branched into long main root (Mr) and short accessory root (Ar). The main root is extended to the basal portion, at which the nucleus of the columnar cell is located. Bar: 200 nm.

Fig. 3-8 A cross section on the branching level of the main (Mr) and accessory roots (Ar) shows fixed angle of the branching. Bar: 200 nm.

Fig. 3-9 Cilia (C) are continuous to kinetosome (K) and main root. (Mr). The main roots are angled at about 135° to the ciliary extension. Bar: 200 nm.

Fig. 3-10 Longitudinal section exhibits the connecting fashion of the cilia (C) to the kinetosome (K) and main root (Mr). Bar: 200 nm.

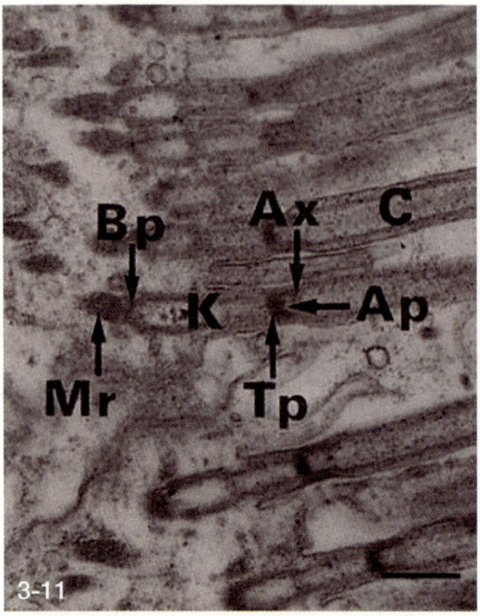

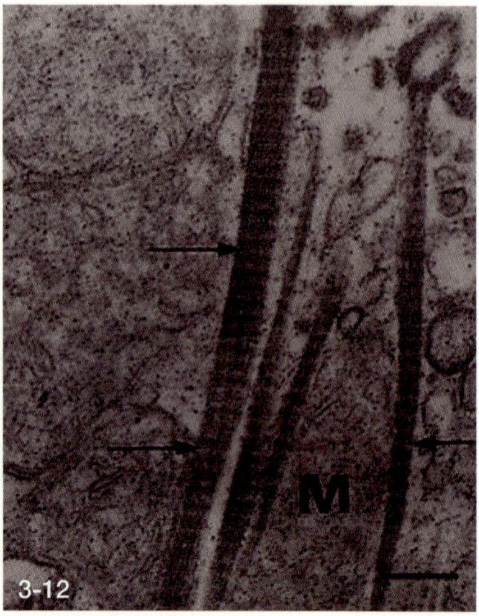

Fig. 3-11 At cilium beneath, the microtubules of the cilia are fused one another to form axosome (Ax), axosomal plate (Ap) and terminal plate (Tp). The terminal plate is continuous spatially with the kinetosome which is terminated at the basal plate (Bp). Bar: 200 nm.

Fig. 3-12 The 5 nm particles are scattered between the cross striation of the ciliary roots. The appearances of the particle are depended on the fixation condition and related closely with ATPase. Bar: 200 mn.

of a cilia cell, ~200 cilia can be observed on polygonal-free surface (Fig. 3-9). The rest of the kinetosome is embedded in the cytoplasm. The cilium (16~18 μm) runs from the axosome of the kinetosome and extends past the cell membrane.

The kinetosome inclines towards the direction of ciliary movements while the main root is arranged perpendicularly to the surface of the ciliary cell (Fig. 3-10). Beginning from the basal plate of the kinetosome, the main root (16 μm) extends deeply into the endoplasm of the ciliary cell until it reaches the basal nucleus (Shin & Lee, 1998) (Fig. 3-10, 3-12). Numerous mitochondria of various sizes (500 nm~800 nm in diameter) are aligned along the main root of the cilia. The main root is in close proximity with mitochondria that are arranged along the root in a column. Sometimes, the cross-striations of the main root approach closely the neighboring mitochondria.

Short accessory roots (400~500 nm in length) extend obliquely from the main root at the basal plate beneath (Fig. 3-7~ 3-8). The accessory root grows in the opposite direction of the ciliary inclination at an oblique 80° angle to the main root (Fig. 3-7, 3-8). Cross-striations are observed on the accessory roots. Cross-striations of the ciliary roots are more clearly seen in the upper portion of the roots with repeating patterns of 20 nm dark and 74 nm light bands appearing along the entire length of the root (Fig. 12).

Numerous particles 5~7 nm in size are dispersed in the cross-striation of the ciliary root and mitochondrial column of the main root, particularly concentrated near the cross-striations of the main roots. The appearances of these particles and striations are related to what fixatives are used. The particles and striations can be seen when the cells are fixed in 2.5% glutaraldehyde. When 4% glutaraldehyde is used, fewer particles and striation are observed.

3-1-3. Ciliary function of the filtration apparatus

Kinocilia are a differentiated form of cilia, and can be distinguished from stereocilia based on structural differences. Kinocilia are consisted with 9+2 microtubule arrangement (Garey & Riggs, 1984; Hiller 1993b), develop main root (Kawamura & Maruyama, 1991) and accessory root in cytoplasm of cilia cell (Krstic 1976, 1982). On the other hand, the kinocilia are divided into two groups according to presence of the accessory root. Distinguished from the kinocilia with accessory root, the kinocilia without accessory root are found in cilia exposed in competitive environment for food materials (Best & Thorpe, 1983, 1986). The accessory root of the cilium is not frequently found, especially in the inner ciliary organ of metazoan except for Bryozoa (Best & Thorpe, 1983). An accessory root is observed often in internal filter apparatuses in which cilia move food materials in the absence of competition (Strathmann, 1982). *Urechis* filtration apparatus is an internal organ and the kinocilia are not exposed in competitive environment gathering foods from the inflows of sea water. It is considered that the ciliary movement in support of accessory root is effective to select planktonic foods rather than indiscriminative gathering including non-food materials. After the infiltration, an amount of sand and shell particles are discharged together with outflow of sea water. The ciliary movements selecting the planktonic foods seem elaborated comparing with the ciliary movement of external organ.

The microtubules of the cilium are continuous with the kinetosome to the main root and accessory root. The microtubules fuse with one another at the axosome or axosomal plate. Beginning from the base of the axosome, kinetososmal fibers pass though the kinetosomal hole and fuse together at the basal plate (Rudman, 1972). Development of two successive plates or lamellar structures in the kinetosome implies that the kinetososme is in function of the start point of the ciliary movement rather than serving directly on the ciliary movement itself (Wehland *et al.*, 1983; Shin 1998). The main root extends downward from the base of the basal plate perpendicular to the tangential line of the cell surface regardless of the kinetosomal inclination. The arrangement of the main root implies that it plays a role in fixing the cilia to promote ciliary movement.

Cross-striation of the ciliary roots are consisted with tubulin (Eisler, 1988), a protein involved in ciliary movement. On the other hand, particles (5~7 nm in size) scattered all over the main root are reported to influence the efficiency of the ciliary movements. The particles are concentrated in the cross-striations, thought as precursor of tubulin molecules in *Ciliophora* and *Copepoda* (Aescht, *et al.,* 1991; Hiller, 1993a; Raff *et al.*, 1997), or ATP aggregates (Dentler, 1988; Wehland *et al.*, 1983; Rodionov & Borisy, 1985). The latter hypothesis is based on the location of the particles and their acute sensitivity to aldehyde fixatives. In *Urechis*, these particles and cross-striation are sensitive to higher concentrations of aldehyde fixatives. Few particles are observed when the cells are fixed in fixative containing 4% (v/v) glutaraldehyde.

The inflows and outflows of sea water are repetitive as the esophageal walls are expanded and contracted. But the muscle layers of the esophagus or foregut are not well developed to initiate expansion and contraction of the esophagus. The in- and outflows of sea water and gathering of foods is more effective with the aid of the body wall containing myosin molecules (Magnum *et al.*, 1983). Positive and negative pressures are created in the esophageal lumen by repetitive expansion and contraction of the body wall. When the body wall expands, negative pressure is transmitted by the numerous ligaments onto the esophageal surface and leads to the expansion of the esophagus. Sea water flows into the esophageal lumen, and cilia move planktonic foods towards the opening of the digestive tract at the end of the esophagus. With the continuous and gentle application of positive pressure, sea water flows slowly out of the esophagus. The lumen of the esophagus gradually narrows and food particles become trapped between the entangled cilia. In addition, the negative and positive pressures created by the body wall cause sea water to flow into the hind gut. In- and out-flowing sea water have different pO_2 and pCO_2 which are closely associated with the respiratory activity of the worm (Garey & Riggs, 1984; Hall *et al.,* 1981; Magnum *et al.*, 1983).

Summary

Filtration apparatus of *Urechis* is anatomized under LM and EM. The unbranched digestive tract can be divided into three sections according to function. The foregut functions as filtration apparatus, the midgut is the actual digestive organ, and the hindgut serves as a respiratory organ. Tissue organization of the filter apparatus is pseudostratified ciliated columnar epithelium composed with three types of cells: numerous ciliary cells, mucous and other kind of secretory cells.

Cilia is measure 16~18 μm long, differentiated in structure with kinetosome, accessory and main root which extend deeply into cytoplasm of the ciliary cell. Fine structures of kinetosome and striated ciliary roots are descried. Numerous cilia covering over the inside surface of the esophagus move food materials contained in the absorbed sea water to the bottom of the esophageal orifice. Ciliary beating alone is not enough to filter food materials. Negative and positive pressures created by muscular movements are also required filtering. The inflows and outflows of sea water are repetitive as the wall of foregut is expanded and contracted. Negative pressure is transmitted by the numerous ligaments onto the foregut surface and leads to the expansion of the esophagus. Inflow of sea water is in common rapid followed by gently and slow outflow. The foregut lumen gradually narrows and food particles become trapped and filtered between the entangled cilia.

Filtration of *Urechis* is differed from filtration of other filter feeder and characterized as follows: first, filter apparatus is an internal organ, a part of digestive lineup. Second, filtration is coordinated with muscular actions of body wall creating positive and negative pressures. Third, the cilia are well differentiated in structure developing kinetosome and accessory root, which is lack in external ciliary apparatus of other filter feeder.

Negative and positive pressure created separately at posterior body wall cause sea water to flow into the hind gut is closely associated with the respiratory activity of the worm.

REFERENCES

Aescht E, W Foissner, M Mulisch: Ultrastructure of the Mycophagous ciliate *Grossglockneria acuta* (*Ciliophora, Copodea*) and phylogenetic affinities of colpoid ciliates. Europ J Protistol, 26: 350-364, 1991.

Akowska JJ, EM Nelson, J Frankel: Development of the ciliary pattern of the oral apparatus of *Tetrahymena thermopila.* J Protozool, 29(1): 366-382, 1982.

Austin P, F White: Metabolism and oxygen transport in the Innkeeper *Urechis caupo.* Physiol Zool, 54(1): 44-54, 1981.

Best MA, JP Thorpe: Effects of particle concentration on clearance rate and feeding current velocity in the marine Bryozoan *Flustrellidra hispida*. Marine Biol, 77: 85-92, 1983.

Best MA, JP Thorpe: Effects of food paricle concentration on feeding current velocity in six species of marine Bryozoa. Marine Biol, 93: 255-262, 1986.

Bradbury PC, LS Olive: Fine structure of the feeding stage of a sorogenic ciliate, *Sorogena stoianovitchae* gen. n., sp. n. Protozool. 27(3): 267-277, 1980.

Dentler WL: Fractionation of Tetrahymena ciliary membrane with Triton X-114 and the identification of a ciliary membrane ATPase. J Cell Biol, 107(6): 2679-2688, 1988.

Edmond SJ, AC Stephan: The phyla Stomata and Echiura. In "Invertebrate Zoology". p.465 ff R.D. Barnes ed., Saunders College Publishing, 1972.

Eisler K: Electron microscopical observations on the ciliate Furgasonia. Blochmanni Faure-Fremiet. Europ J Protistol, 24: 75-93, 1988.

Garey JR, AF Riggs: Structure and function of hemoglobin from *Urechis caupo*. Arch Biochem Biophys, 228: 320-331, 1984.

Hall RE, RC Terwilliger, NB, Terwilliger: Hemoglobins and myoglobin of the Echiuran *Urechis caupo* (Fischer and Macginite). Comp Biochem Physiol, 70: 353-357,1981.

Hannes A, W Foissner: Morphologie und oekologie einiger Ciliaten (Protozoa: Ciliophora) aus dem Belebtschlamm. Arch Protistenkd, 141: 243-283, 1992.

Hiller SA: Ultrastructure of prorodon (*Ciliophora, Prostomatida*). Ⅰ. Somatic cortex and some implications concerning kinetid evolution in Prostomatid and Colpodid Cilisate. J Europ Microbiol, 40(4): 467-486, 1993a.

Hiller SA: Ultrastructure of prorodon (*Ciliophora, Prostomatida*). Ⅱ. Oral cortex and phylogenetic conclusions. J Europ Microbiol, 40(4): 486-501, 1993b.

Hofmann-Muenz AH: The Oral apparatus of colpoda variabilis (*Cil*iophora, Colpodidae). Ⅱ. Ultrastructure of the oral ciliature and its implications on ciliate phylogeny. Europ J Protistol, 26: 288-302, 1991.

Kawamura NYK, K Maruyama: Myosin from body wall muscle of Annelid, *Urechis unicinctus*.

Zool Sci, 8(6): 1130-1142, 1991.
Krstic RV: Ultrastucture der Saegetier zelle. Ein Atlas zum Sudium fuer Mediziner und Biologen. 350pp. Springer verl, Berlin-Heidelberg- New york, 1976.
Krstic RV: Die Gewebe des Menschen und der Saegetiere. Ein Atlas zum Studium fuer Mediziner und Biologen. 18pp. Springer verl., Berlin-Heidelberg- New york, 1982.
Lynn D: Implications of recent descriptions of kinetid structure and systematics of the ciliated protists. Protoplasma,164: 123-142, 1991.
Mangum P, RC Terwilliger, NB Terwilliger, R Hall: Oxygen binding of intact cells and extracted hemoglobin of the Echiuran Urechis caupo. Comp Biochem Physiol 76A(2): 253-257, 1983.
Raff EC, JD Frankentha, JA Hutchens, HD Hoyle FD Tumer: Microtubule architecture specified by a β-tubulin. Science, 275: 70-73, 1997.
Remana A, V Storch, U Weisch: Systematische Zoologie. Staemme Tierreichs. 28pp. Gustav Fischer Verl. Stuttgart, 1975.
Rodionov VI, GG Borisy: Microtubule treadmilling *in vivo*. Science, 275: 215-218, 1997.
Rolf S: Echiurida In "Lehrbuch der Zoologie" Band 2, Systematik, 3 rd ed., p574 ff.,Gustav Fischer Verlag. Stuttgasrt-New york, 1985.
Rudman WB: Structure and functioning of the gut in the Bullomorpha (*Ophistobranchia*). J Nat Hist, 6: 547-560, 1972.
Shin KS: The fine structure of sperm ball and sperm of *Urechis unicinctus* and immunogold localization of α-tubulin. Korean J Electron Microscopy. 28(2): 193-205, 1998.
Shin KS, SH Lee: Ultrastructure and filtrating function of the ciliated epithelial cells of foregut in *Urechi unicinctus*. Korean J Electron Microscopy, 28: 435-446, 1998.
Strathmann RR: Cinefilms of particle capture by an induced local change of beat of lateral cilia of a Bryozoan. J Exp Mar Biol Ecol, 62: 225-236, 1982.
Wehland J, J Mark, C Willingham, VI Sandoval: A rat monoclonal antibody reacting specifically with the tyrosylated form of α-tububline. I. Biochemical characterization: Effects of microtubule polymerization in vitro and microtubule polymerization and organization in vivo. J Cell Biol, 97: 1467-1475, 1983.

Chapter 04

Spermatogenesis of *Urechis unicinctus*

Chapter 04
Spermatogenesis of *Urechis unicinctus*

4-1 Introduction

Since the beginning of the last century, it has been known that the process of gametogenesis can be divided into localized and diffuse gametogenesis. Additionally, either localized or diffuse gametogenesis selectively proceeds in all animal species (Korschelt & Heider, 1902). In other word, once a pattern is selected by a species, the same gametogenesis process occurs in both males and females. In bisexual reproduction, no exceptional cases have been found to date.

Localized gametogenesis occurs in a wide variety of vertebrates and many species of invertebrates. The primordial germ cells are derived from gonads, and are organized in special organs composed of somatic tissues and germ cells. Organs composed of the compound tissues include testes and ovaries. Primordial germ cells develop and differentiate into mature sex cells with the help of associated somatic cells. Since gametogenesis occurs in distinctive organ in a fixed position, this pattern of gametogenesis is known as "localized" gametogenesis.

In contrast, diffuse gametogenesis is found predominantly in invertebrate species. This process is first observed in female *G. gigas* (Mayer, 1928). Primordial germ cells are enclosed within an "ovarian ball" composed of somatic cells enclosing the primordial germ cells (Mayer, 1928, 1932-1933). Ovarian balls arise from fragments of a single ovary during an early developmental phase (Mayer, 1932-1933). Over time, numerous ovarian balls are scattered and float in the body fluid. Oogenesis proceeds in worms without a fixed position or organized organ. The existence of ovarian balls has been successively confirmed, and the term "diffuse oogenesis" is used when describing this process in various species (*Acanthocephalan;* Awachie. 1965, 1966; Crompton, 1970; Crompton & Whitefield, 1974; Atkinson & Byram, 1976; Shin 1986; *Polychate, P. dumerilli:* Fischer, 1975).

Urechis is an invertebrate species in which diffuse gametogenesis occurs. Several authors have reported the details of diffuse oogenesis (Gould-Somero, 1975; Sawada & Ochi, 1962: Paul, 1975; Gould-Somero & Hollander, 1975b; Shin, 1998; Kwon *et al*., 2003). However, oogenesis in *Urechis* differs greatly from typical diffuse oogenesis in two ways. First, diffuse units in the coelomic fluid are not associated with somatic cell, but are solitary during the entire development process until they are discharged for fertilization (Jeong, 1989; Kwon, 2003; Kwon *et al,* 2000). Germ cell aggregations are found early during proliferation, but there no associated somatic cells are observed in the aggregates. At the next stage, individual germ cells are dissociated from the aggregates. Thereafter, individual germ cells exist as solitary cells until they are discharged for fertilization.

Second, the developing sex cells are observed in the storage sac, a localized organ. There are differences of opinions among authors about the precise function of the storage sac. Some believe that the storage sacs are the organs responsible for storing mature sex cells (Gould-Somero, 1975; Sawada & Ochi, 1975; Paul, 1975; Gould-Somero & Hollander, 1975a). Others reported that the storage sacs are another sex organ in which germ cells are generated and developed (Jeong 1989: Shin 1998). These latter authors observed developing spermatocytes enclosed in a "sperm ball", a mass of somatic cell enclosing a number of male germ cells. Even though insufficient evidence has accumulated and the observed stage is limited to a special stage, data indicating that the storage sacs is a type of sex organ comes from the observation of spermatids in sections of the *Urechis* storage sac (Jeong, 1989). The existence of a sperm ball containing spermatids has been confirmed in a subsequent study (Shin 1998).

Conceptually, both localized and diffuse gametogenesis appear to occur in *Urechis* species even though the stages and stage distribution of localized spermatogenesis have not yet been established in detail. Additionally, developing male gametes are found as above in the male storage sacs, but there is no available data to prove the developing female gametes in the female storage sacs.

4-2 Spermatogenesis in the male storage sac

When worms are prepared between February and middle April, the upper or proximal part of the male storage sac (MSS) is milky white in color, and partial diminutive expansions can be seen at various locations along its length. The cross section through the expanded sites shows that the MSS wall is thinner and has a widened lumen in which mature sperm accumulate (Fig. 4-2). In contrast, the lower distal parts of the MSS are brown in color, and slender with thicker MSS walls and narrow lumen. The MSS wall is thick and contains stratified layers of accumulated cells (Fig. 4-1). No male gametes are found in the distal portion of the MSS. Developing male gametes can be seen on sections prepared from the interfaces between the proximal and distal MSS. In the lumen of the MSS, the male gametes are enclosed in a "sperm ball" (Fig. 4-3), a separate structural unit consisting of central somatic cell and developing male gametes located mainly at the marginal sites of the sperm ball (Jeong, 1989; Shin, 1998).

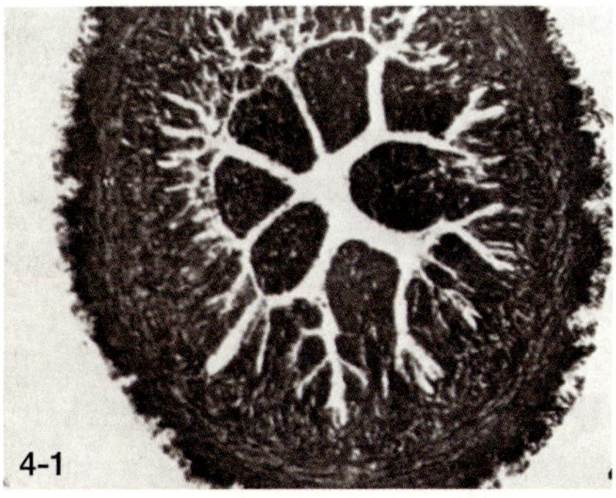

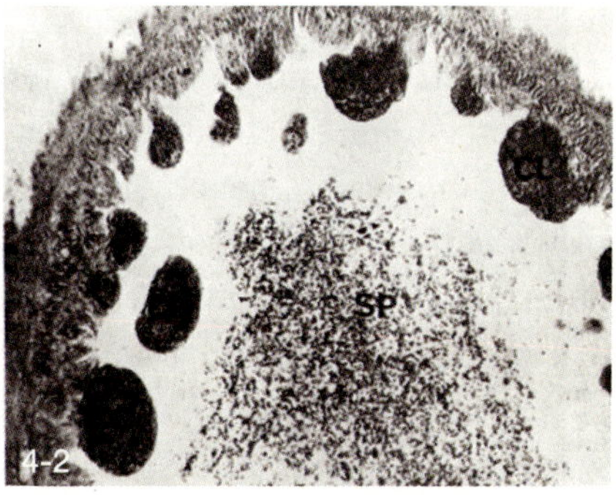

Fig. 4-1 Prepared at earlier breeding season, lower distal parts of the male storages sac are slender with thick MSS wall. No mature sperm can be observed in middle lumen. × 400.

Fig. 4-2 Upper proximal parts of male storage sac (MSS) show the thin MSS wall, include mature sperm (Sp) in the middle lumen of the MSS. Compared with the distal part (Fig. 4-1), sperm ball consisted with somatic cell and male gametes seem detached from the MSS wall. × 400.

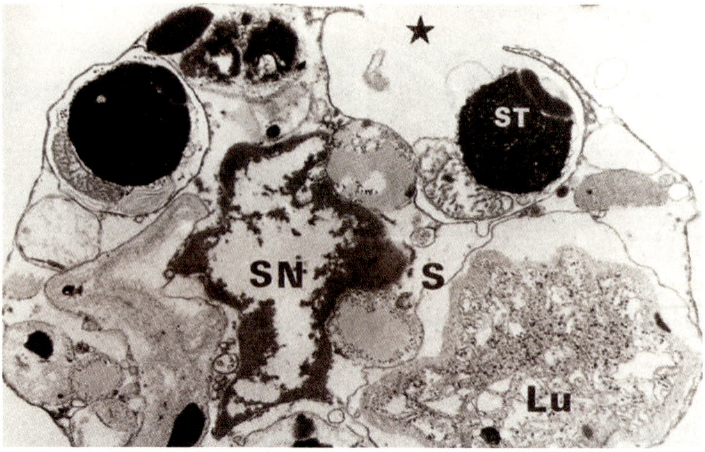

Fig. 4-3 Sperm ball prepared from interfaces of distal and proximal MSS is consisted with somatic cell (S) and spermatids (ST) arranged along cortical cytoplasm of the somatic cell. A perispermatocyte space containing spermatid is enclosed, other is opened and continuous with lumen (asterisk) of the MSS, and another (Lu) reveals accumulations of disordered material which may remains of spermatid detachment from the perispermatocyte space. SN: Nucleus of somatic cell. × 6,000.

It has been noted that developing male gametes observed in the sperm ball are mainly spermatids (Jeong, 1989; Shin, 1998). Other developmental stages exclusive of spermatids are not found in the literature, and should be studied to understand the spermatogenesis process in the MSS. Spermatids are enclosed in a large sperm ball (10~30 μm × 8~15 μm) which contain somatic cells and several developing spermatids at the marginal cytoplasm. The nucleus (4~5 μm × 3~4 μm) of the somatic cell includes a large amount of heterochromatin and the nuclear envelope is heavily undulated. The cytoplasm is characterized by abundant lipid droplets. Spermatids are found in the perispermatocyte space (3~5 μm), which is a cytoplasmic expanse delineated by cell membranes of the somatic cells.

The flagellated spermatids in the sperm ball contain condensed nucleus (2 μm in diameter) with a small amount of cytoplasm and organelles: acrosome in the anterior nuclear fossa as well as mitochondria and centrioles in the posterior nuclear fossa. The general morphology of the spermatids is identical to spermatid enclosed in the late stage spermatid cluster (LSSC) in coelomic fluid. The membrane of the somatic cell enclosing the perispermatocyte space exhibits various contour lines. Some perispermatocyte spaces are intact and completely enclose the spermatids. Others are partially opened at the marginal site and continue with the MSS lumen. Finally, others are empty without any spermatids and have a wide opening into the MSS lumen.

The programmed movement of spermatids and coupling of the contour lines form an inference that the spermatids develop in the perispermatocyte spaces. Advanced spermatids are released through the marginal opening into the MSS lumen, leaving the space empty. Often, cytoplasmic debris accumulates in the empty space.

Oblique sections through the border line between the proximal and distal MSS provide reliable evidence that sperm balls are derived from cell components detached from the MSS wall (Jeong, 1989, Fig. 4-1, 4-2). The wall of the distal MSS is over 30 μm thick with an undulating inner surface consisting of several cell layers folded over each other. The thicker wall of the MSS causes the lumen to be narrowed. Over time, the lumen is widened while the MSS wall becomes much thinner due to the successive detachments of the cell layers from the MSS wall.

4-3 Spermatogenesis in the coelomic cavity

Preparations of male coelomic fluid made between February and March contained male gametes at various developmental stages (Shin & Kim, 2002). Unlike spermatogenesis in the MSS, spermatogenesis in coelomic fluid is carried out in a diffused manner without tissue organization. Studies of diffuse spermatogenesis offer more details about developmental stages and stage distributions of male gametes.

At the initial appearances, several spermatogonia aggregate together and float in the coelomic fluid (Fig. 4-4). The aggregates seem to be derived from successive divisions of an archetype cell (Shin & Kim, 2002). The aggregations are unique and easily recognizable at the right time of the sample preparation. Individual spermatogonium enclosed in the spermatogonia aggregate can be identified under light microscopy and appear different from coelomocytes in several aspects. The spermatogonium is smaller (6 μm

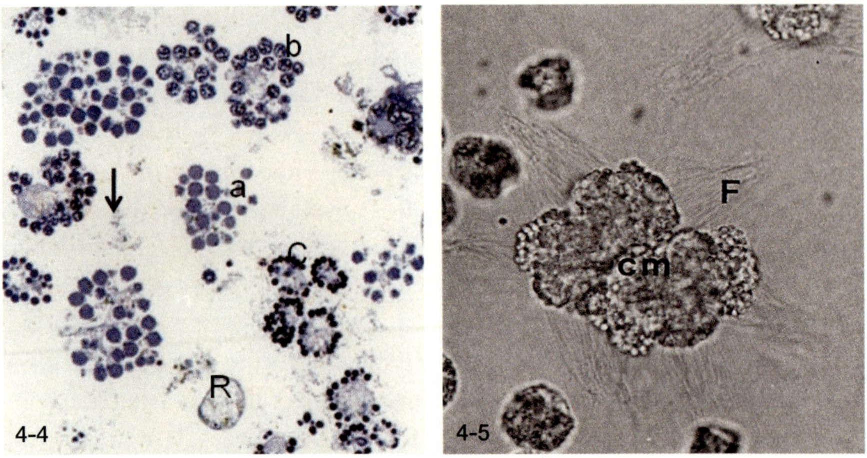

Fig. 4-4 Various developmental stages of spermatids clusters are observed from preparation of coelomic fluid. Arrow indicates entangled mature sperm, and early stage of spermatid cluster (a), intermediate stage (b) and late or final stage of spermatid development (C) as well as residual body (R) are observed from the preparation of coelomic fluid. × 200.
Fig. 4-5 Numerous developing spermatids are attached to cytoplasmic mass (CM). Long flagella (F) of the spermatids are bundled up extending outward the mass. × 1,200.

in diameter) than the coelomocyte, semitransparent, slightly ellipsoid in shape with an oval nucleus (3 × 4 μm) containing many heterochromatin particles in the nucleoplasm. The coelomocyte is always alone and do not associate with each other.

At the subsequent stages, developments and growths of the individual spermatogonia continue in the aggregate which developed into spermatid cluster (Fig. 4-4, 4-5). After formation of the syncytial spematid cluster, spermiogenesis is progressed in the spermatid cluster and mature sperm are detached finally from the spermatid cluster into coelomic fluid. The processes of the spermatid cluster formation and spermiogenesis is characteristic, subdivided into three stages: early, intermediate, and late stages.

4-3-1. The spermatids in the early stage of spermtid cluster

In mid-February, the male gametes in the aggregate are changed into rather larger and oval-shaped cells 6 × 8 μm in size with massive nuclei measuring 5 × 6 μm. Although the oval contour of cell and large size of the nucleus, the male gametes seem enter the spermiogenesis by the reason of the cell organelles of spermatid can be observed in the cytoplasm. Acrosomal, centrioles and flagella precursors have been observed in the oval cytoplasm of the primary spermatid (Fig. 4-4, 4-6).

At the stage of the primary spermatid, the most characteristic features can be observed at the contact site with the neighboring primary spermatocytes. Although the individual spermatid is discernible on the whole and distinguished from the other spermatids, the cell membrane is fused with other cell membrane of the neighboring spermatids. On the other hand, the elongated vacuoles are arranged along the nuclear envelope of the primary spermatocytes leaving an impression behind that the nucleus is isolated from the cytoplasm.

The continuations of the cell membrane fusion bring the primary spermatids into spermatid syncytium at next intermediate stage of spermatid cluster (ISSC). At the beginning of the syncytium formation, the spermatid nucleus with large size (5 × 6 μm) and homogeneous nucleoplasm is maintained until the fused site of the plasma membranes developed into a cyto-

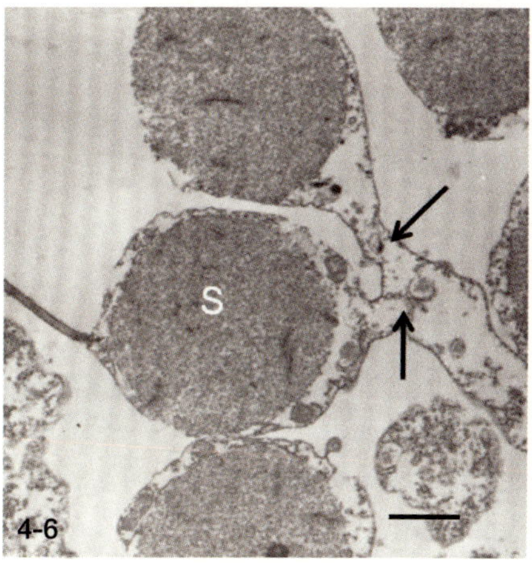

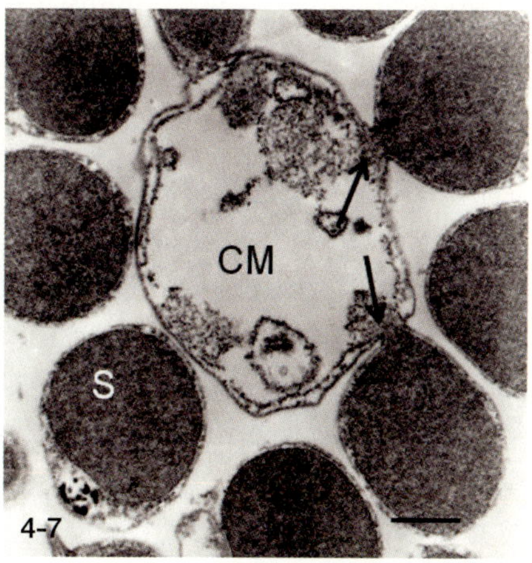

Fig. 4-6 It is likely membrane of individual spermatocyte fused each other to form spermatid syncytium, spermatid cluster. Male gametes of *Urechis* developed asynchronously, some of spermatocytes have been developed into spermatid possessing flagella. Arrows indicate sites of close contact and membrane fusion. Somatic cell is not observed in spermatid cluster. Bar: 1 μm.
Fig. 4-7 An earlier stage of spermatid cluster (ESSC) shows relative large nucleus and decreased amount of spermatid cytoplasm. Cytoplasm of marginal spermatid flows into central cytoplasmic mass through cytoplasmic connection (arrows). Bar: 1 μm.

plasmic connection which looks like a cytoplasmic bridge between the cells. In advance of time, the fused cytoplasm derived a number of the primary spermatids is developed into central cytoplasmic mass, and the spermatids are arranged along the marginal sites of the common central cytoplasmic mass of the spermatids. The spermatids are attached to the cytoplasmic mass by narrow cytoplasmic connections. Over time, a reduction in amount of spermatid cytoplasm is clearly observed while the amount of the cytoplasmic components of the central mass increase. The cytoplasmic components of the central mass are not organized; figures resembling disorganized ER, Golgi bodies, and other membranous debris can be observed in the cytoplasmic mass.

During the subsequent ESSC stage, the nuclei of the marginal spermatids decrease in size (3 × 4 μm in diameter), and the volume is reduced up to 27% of the primary spermatid nucleus (Fig. 4-7). Compared to primary spermatids, the external shape of the latter spermatid nuclei are not changed except the increased density of the nuclear matrix. The acrosomal precursor, mitochondria, and centrioles are remained, while other cytoplasmic components are not to be seen in the spermatid cytoplasm. The spermatid organelles are not randomly distributed in the spermatid cytoplasm, they tend to be located at the site opposite to the cytoplasmic connection.

4-3-2. The spermatids in the intermediate stage of spermatid cluster

The ISSC is differed from the ESSC in plenty of the central cytoplasmic mass and condensed nuclei of the marginal spermatids (Fig. 4-8). The ISSC spermatid nuclei are measured 2.5~3 μm in diameter, reached about 12% of primary spermatid nuclei. The nuclear volume of the ISSC spermatid is reduced by about 88%. On the other hand, the nuclear density is greatly increased and nuclear vacuoles appear in the nuclear matrices. Along with reduction of the nuclear volume, the appearance of nuclear concavity is another morphological evidence of the nuclear condensation. Consequently, the oval nuclei of the ESSC spermatids is transformed into a round and compact form with the concavity at the prospective acrosomal site.

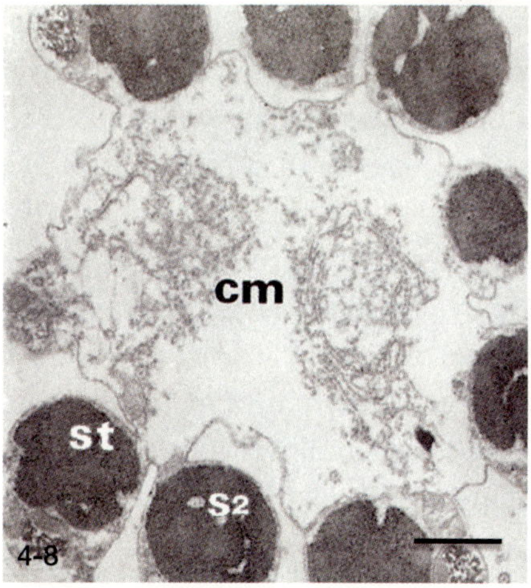

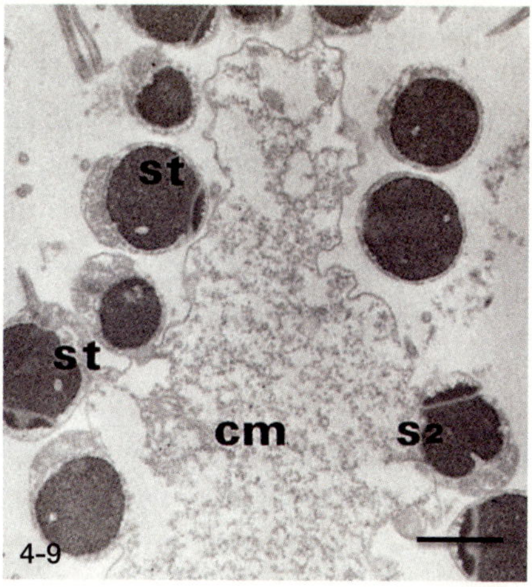

Fig. 4-8 Intermediate stage of spermatid cluster is recognized by number of large nuclear vacuoles and granular nuclear matrixes. Development of spermatids is asynchronous, and advanced spermatid (St) and retreated spermatid (S2) are enclosed in the intermediate spermatid cluster. cm: Central cytoplasmic mass. Bar: 1 μm.

Fig. 4-9 Spermatid nuclei are more compacted and condensed, in possess of small and less number of nuclear vacuoles in late stage of spermatid cluster (LSSC). Advance spermatid (St) and retreated spermatid (S2) in development are to be observed in the LSSC. cm: Central cytoplasmic mass. Bar: 1 μm.

The central cytoplasmic mass is also greatly increased. Some spermatid organelles, such as acrosomal precursors, centrioles, and flagella, congregate together at the site opposite of the cytoplasmic connection. The site opposites seem temporary site of the spermatid organelles, since they are changed its position at next developmental stages. For instance, once settled at the site opposite to the cytoplasmic connection, the acrosomal precursors (1.2 μm × 700 nm) move again toward the lateral site of the spermatid.

4-3-3. The spermatids in the late stage of spermatid cluster

LSSC spermatids are different from spermatids in previous clusters, and can be recognized at a glance by the presence of a bipartite acrosome (Fig. 4-9). The number of nuclear vacuoles is appeared in the nucleoplasm, and the nuclear matrices are rather densely packed and granulated. Another characteristic of LSSC spermatids is that the abundant cytoplasm is accumulated in the central cytoplasmic mass.

The spherical nuclei of ISSC spermatids are transformed into round disc forms (1 μm in thickness × 2 μm in diameter) with prominent frontal concavity, at which acrosome is settled in future. The nuclear volume is estimated at 2% of that of primary spermatid nuclei. In proceeding of the LSSC development, the spermatid organelles are migrated and settled down at the proper sites of the cell organelles. The acrosome is located at the anterior nuclear fossa as described above, while mitochondria and centrioles adhere at the posterior nuclear fossa.

At the anterior nuclear fosse, the acrosome finally differentiates into a round disc form (1.5 μm × 400 nm) with a central hollow (200 nm in diameter). Bipartite gradations of acrosomal materials are also established (Fig. 5-6, 5-7, 5-9). The outer circumferential regions are electron dense in contrast to the less dense inner regions. The entire acrosome is enclosed within a membranous structure, but no membrane structures are observed at the boundary of the bipartitions. The circular band-like shape of mitochondrion is characteristic. The mitochondrion is arranged perpendicular to the spermatid head axis, and the proximal and distal centrioles are positioned at the region intersecting the spermatid head axis and mitochondrial plane. Cytoplasmic connections between the LSSC spermatids and central cytoplasmic mass are narrow. No structural devices have been observed in these connections. Vacuoles with irregular forms are often seen in the cytoplasmic connections. The vacuole seems to be ultimately dedicated to detaching spermatids from the LSSC. Cytologically, spermatid detachment can be compared to the disjunction of exocytotic vesicles from somatic secretory cells.

The functions and shapes of the cytoplasmic mass resemble residual body known in spermatogenesis of the localized spermatogenesis.

Anuclear cytoplasmic masses correspond to synplasm derived from successive endomitosis accompanying delayed cytokinesis. Observations over time have raised a possibility that the cytoplasmic mass plays a role in eliminating excessive cytoplasm from developing spermatids.

4-3-4. Mature sperm

After detaching from the LSSC, mature sperm does not belong to the spermatid cluster, but exists as individual cells in the body fluid. Besides this characteristic, mature sperm are different from less developed spermatids in the LSSC because of their condensed nuclei with low amounts of cytoplasm and complete differentiation of cellular organelles (Fig. 5-11).

The axis of a mature sperm head is aligned at an 110° angle to the flagella axis. The shape of the sperm nuclei is a compact and round disk (1×2 μm) with anterior concavity. The nuclear vacuoles and extra space between granular nuclear matrices disappear with a few exceptions. The nuclear volume and size seem not greatley decreased compared to LSSC spermatid. Without an obvious volume condensation, the granular nuclear matrix of LSSC spermatids is homogenized and stabilized in the course of development into mature spermsatozoon.

The acrosomes of mature sperm are highly differentiated an d have an elaborated shape compared to those of other marine animals. The massive acrosomes (1.5 μm × 500 nm) are shaped as elliptically concave discs, similar to an elongated concave lens, and contain a central hollow (200 nm in diameter). The acrosomes are located in the unconstrained anterior nuclear fossa at forefront of the perinuclear theca. The circumferential masses of the central hollow are bipartite and contain differential gradations of material concentrations. The outer mass is denser than the inner mass. Bipartition of the acrosomal materials is necessary for certain functions performed during acrosomal reactions (Shin *et al.,* 2000). The inner acrosomal materials are used to form the prospective acrosomal sheath, and the outer material is dispersed and adheres to the spermatozoon on oocyte (Kwon *et al*., 2000).

Although much of the cytoplasm is eliminated during the LSSC stage, some of the cytoplasm is retained in mature sperm, especially in the posterior nuclear fossa, in which circular mitochondrion and centrioles are located. The proximal and distal centrioles are aligned around the center region of the circular mitochondria. The proximal centriole is directly connected to fractional extension of the nuclear envelope, and the distal centriole connects to the frontal tip of the flagellum. The mitochondrion of mature sperm is ring or circular band in shape derived from close contact of three mitochondria.

The border line of the mitochondrial contacts can be observed under EM.

Without fibrous sheath, the mitochondrial arrangements and shape are particular features greatly differ from disjoined and arbitrarily ordered mitochondria in other animals.

4-4 Localized and diffuse spermatogenesis

As seen above, it can be concluded that *Urechis* sperm arise from two types of spermatogenesis: the non-organogenic diffuse spermatogenesis that occurs in the body cavity (Gould-Somero, 1975; Gould-Somero & Hollander, 1975a; Sawada & Ochio, 1975: Paul, 1975; Shin 1998), or organogenic localized spermatogensis that takes place in the male storage sac (MSS) (Jeong *et al.,* 1989; Shin, 1998). Mature sperm produced in the coelomic fluid are called "coelomic sperm" in order to distinguish them from "MSS sperm" which are derived from sperm balls in the MSS. Since the sperm derived from both processes have equal fertilization capacities, these terms are not based on function of the sperm, but are descriptive names indicating their separate developmental origins.

Even the observation of MSS sperm have been limited to spermatid stage enclosed in the sperm ball and the developmental stages of localized spermatogenesis have not been precisely determined, developing male gametes and localized spermatogenesis seem to proceed in level of special organ. To date, no previous publications have described the two patterns of spermatogenesis.

Artificial fertilization experiments have shown that mature sperm derived from both processes have equal fertilization capacities (Choe, *et al.*, 1988; Shin, *et al.*, 2000) from which the fertilized egg develops into a feeding larva, trocophore (Shin *et al.,* 2004). As above, the localized spermatogenesis have not been enough analyzed, it is preferable to describe the diffuse spermatogenesis in detail. The common source of *Urechis* spermatogonia should be further investigated.

4-4-1. Diffuse spermatogenesis in *U. unicinctus*

Observations of diffuse spermatogenesis include the followings. Coelomic spermatogonia first appear as aggregates floating in the coelomic cavity. Belonging to the aggregate, the development and growth of the spermatogonia continue until reached individual spermatocytes. In the spermatocyte with large nucleus and plenty of cytoplasm, precursors of the acrosome, centrioles and flagellum are appeared. At next stage, the spermatocytes develop into spermatids enclosing in early stage of spermatid cluster (ESSC). The spermatid cluster of *Urechis* is a syncytium in structural aspects consisted with central cytoplasmic mass and spermatids linked with cytoplasmic connections along the marginal site. Although the reason and cause of the spermatid cluster formation is not yet probable, there are two possible viewpoints for the formation of the syncytium: one is rapid nuclear division and delayed cytokinesis of the single spermatocyte, and the other is fusion of cytoplasm of the spermatocytes. The evidence for the formation of the syncytium is not available. Shin & Kim (2002) observed the developments of the spermatocytes into spermatids and described that the individual spermatocytes are fused one another in the process of differentiation into spermatid stage.

Further development of the spermatids is continued in the spermatid cluster until completion of the spermiogenesis. Hence, spermatid clusters are thought to play at least two key roles in spermiogenesis. One is to serve as structural supports, and the other is to eliminate excessive cytoplasm from each spermatid. The removed cytoplasmic mass is equivalent to the residual body of other animals. On the other hand, the sperm heads are greatly condensed during the stages of the spermatid cluster. The LSSC spermatid volume is estimated to be 2% of the primary ESSC spermatid volume.

During spermiogenesis and nuclear condensation, sperm organelles are rearranged in accordance to polarities. At ESSC formation, the spermatid organelles are located at opposite site to the cytoplasmic connection. In advance of time, mitochondria and centrioles are located at the posterior nuclear fossa while the acrosomes migrate at first to lateral site and then to the prospective anterior nuclear fossa of the spermatids. The shapes of the acrosome and mitochondrion are characteristic, not reported previous literature. The acrosomes are elliptical discs with a central hollow. The circumference acrosomal material is bipartite with dense outer and pale inner portion in the material constitution. In artificial fertilization experiments, the outer and inner acrosomal materials exhibit separate function: the outer materials play role in adhesive, while the inner materials dedicate to form acrosomal sheath. The mitochondrion forms a ring or circular band, and arises from the close association of three mitochondria on the same plane arranged perpendicular to the axial line of the spermatid head.

Coelomic sperm are apparently stored in the MSS for prospective spawning. However, it is not known how precisely coelomic sperm accumulate in the MSS. It is likely that the opening at the distal end of the MSS plays an important role in selective collection of mature coelomic sperm. Mature sperm accumulated in the MSS are ejaculated through the sexual pores opened just beneath the proboscis.

4-4-2. Cytoplasmic mass in spermatid cluster

As mentioned above, *Urechis* spermids are enclosed in spermatid clusters and transform into condensed and highly mobilized mature sperm equipped with long flagella. Excessive cytoplasm of the spermatid is eliminated by inward flow into the cytoplasmic mass through narrow cytoplasmic connections. The amount of cytoplasmic mass is inversely proportional to the decreased quantities of spermatid cytoplasm. Numerous solitary cytoplasmic masses without attached spermatid can be seen in preparations of body fluids, especially at the end of the reproductive season. Cytoplasmic masses are considered homologous to residual body of other higher animals according to fate, structure, and function. Residual body is derived from continuous nuclear division or endomitosis followed by delayed cytokinesis. In *Urechis*, it is not clear that the cytoplasmic mass is formed by continuous endomitosis or fusion of the individual spematocytes. Observed evidences in *Urechis* spermatogenesis are preferred to fusion of the individual spermatocyte for formation of the cytoplasmic mass.

Excess cytoplasm remained in the cytoplasmic mass is also eventually discarded. The contents of the cytoplasmic mass include scattered ERs,

Golgi bodies, and cytosol components. Other organelles, which are retained in the spermatids, are not found in the cytoplasmic mass. A selective mechanism separates sperm organelles from the eliminated cytoplasmic components.

Similar kinds of cytoplasmic masses to *Urechis* have been reported in *Annelida* (*Oligochaete*; Jamieson, 1978, 1981), mollusks, and *I. gogoshimense* (Sawada & Ochi, 1962). In the case of *I. gogoshimense*, the cytoplasmic mass is known as "cytophorus" or "anuclear cytoplasmic mass" (Bondi *et al.,* 1976). The syncytiums of *Annelida* are derived from endomitosis of male gametes. The cytophorus and syncytium ensure synchronous growth and development of the male gametes (Sawada & Ochi, 1975; Bondi *et al.,* 1976; MacGuiness & Griswold, 1994). In *Urechis*, the role of the syncytium in inducing synchronous growth is not independent of the cytoplasmic mass. However, increase of cytoplasmic mass in advance of time corresponds with decrease of cytoplasmic components in the developing spermatids. The origin of the cytoplasmic mass is the cytoplasmic components eliminated from developing speramtocytes and spermatids. Enclosed in the spermatid cluster, the most of cytoplasm of the LSSC spermatid is removed except a small portion of cytoplasm in the posterior nuclear fossa with a minimum variety of organelles. Volume of nucleus is reduced into 2% of the primary spermatids found in the earlier ESSC. Although inducement of synchronous growth and developments of spermatids are not exclusive, the cytoplasmic mass in *Urechis* is the outcome of eliminated cytoplasm of the spermatids to make spermatid mobile ultimately.

4-4-3. Nuclear condensation

One of the important events of spermiogenesis is nuclear condensation and nuclear shaping. In most of higher animal species, large spherical spermatocyte nucleus is transformed into a compact and streamlined nucleus to enhance hydrokinetic efficiency in a fluid environment. The nuclear shaping and condensation are cooperative results of developing male gametes with somatic cells, which are associated together within a compact tissue. In *Urechis,* sperm nucleus is compacted into small size, but not shaped into streamlined configuration in the processes of spermiogenesis. In concern with the nuclear transformation, the main events and contents of the spermiogenesis seem decrease of the size rather than shaping of the spermatid nucleus.

It is reveled in animal specis, the cause and effect of nuclear shaping are attributed to intrinsic and extrinsic factors. Extrinsic factors include actin microfilaments of the Sertoli cells and perinuclear manchtte microtubules of the male gametes (Kierszenbaum *et al.,* 2003). The intrinsic factors include nuclear actin microfilaments along with a set of other nuclear proteins (Cooper, 1991; Soyer-Gobillard *et al*., 1996; Kramer & Krawetz, 1996). In *Urechis* sperm, the extrinsic factors are not involved in nuclear shaping and condensation since somatic cells are not associated with developing male gametes during ontogeny. Although another extrinsic factor, perinuclear manchette microtubules, is found in the cytoplasm around the spermatid nucleus, these microtubules are remained as rudimentary microtubules, not organized into a manchette or cylinder shape of microtubules (Shin *et al,* 2005).

Nuclear condensation and shaping of *Urechis* sperm are accomplished solely through intrinsic factors such as the phase-change of the nuclear microfilaments. Repetitive polymerization and depolymerization of nuclear actin are the main factor responsible for nuclear shaping in *Urechis* (Shin *et al,* 2005). Early LSSC spermatids contain large nuclei (5 × 6 μm) with lower nuclear matrix densities. The density of the nuclear matrix gradually increases as the developmental stages progress accompanied by constant reduction of the nuclear size and condensation of the nuclear matrices. The simultaneous size reduction and density increase are morphological evidence of nuclear condensation and shaping caused by repetitive cycles of polymerization and depolymerization of the intrinsic microfilaments as described above (Shin *et al*., 2005). The nuclear condensation in concern with the actin phase changes is revealed in experiments using anti-actin conjugated with gold particle and immunofluorecence methods (*See* Cytological localization of nuclear actin).

4-4-4. Coelomocytes and development of male gametes

In general, the development and strong viability of the sex cells are not possible without structural and function supports of the associated somatic cells (Tres & Kierszenbaum, 1983). This is because some materials derived from somatic cells provide a substantial basis for promoting growth, differentiation, and viability of sex cells. As is described above, the spermatids of *Urechis* are not structurally associated with somatic cells and the spermatid cluster is in consequence a syncytium consisted with the spermatids.

A question arises about how male gametes develop into the functional sperm in the absence of associated accessory cells of somatic origin. As described in the previous section, coelomocytes may play a role in promoting the development of male gametes in *Urechis*. However, sufficient proof has not accumulated in the literature, and no direct evidence is available.

On the other hand, there are a lot of evidence indicating that coelomocyte functions are involved with oocyte development and growth. During oocyte development, some coelomocyte secretions, called "urechrome", enter into the developing oocyte (Hoffmann & Magnum, 1970). Because the pattern of gametogenesis in a species is principally the same in males and females (Korshelt & Heider, 1902), certain substances in the body fluid are required to promote development of *Urechis* sex cells including male gametes. Coelomocytes are unique residents in the coelomic cavity, and share the coelomic cavity with male gametes during the breeding season. Except for coelomocyte, there is no other type of cells that alternate their appearances according to seasonal changes. The ultimate source of the substance that promotes spermatogenic development seems to be coelomocytes. In addition, coincidental increases and decreases in the number of coelomocytes and male gametes and the vigorous secretion of coelomocytes that correspond to the growth phase of the male gametes is evidence of the substantial supportive role of coelomocytes. Additionally, a simple experiment may provide another piece of indirect data indicating that coelomic fluid provides a proper environment for developing male gametes. Mature sperm can survive over 90 hours in the body fluid, but can survive for only 3 minutes in sea water.

4-4-5. Comparative morphology of mature sperm

Sperm are highly differentiated sex cells. Their shape differs from species to species. In developmental biology, the shape of a sperm is often used to identify groups of animals.

Some authors classify *Urechis* as echinoderms because of the similarities in sperm shape. Centrioles and mitochondria embedded in the posterior nuclear fossa with some cytoplasm are in fact similar to those of echinoderms (sea urchin, *Arbacia,* Fawcett, 1972). However, the location and contour of the organelles are not unique features of echinoderm sperm, and can also be seen in mollusks (*Choromytilis meridionalis*, Hodgson & Bernard, 1986) and *Polychaete* (*Neanthes japonica*, Sato & Osanai, 1986).

Acrosomal shapes differ significantly between *Urechis* and echinoderms. *Urechis* arcosomes are massive and shaped as elliptical discs with a central hollow. The hollow is a spare space that is the site of G-actin polymerization during elongation of the acrosomal process. The bipartite acrosomal materials lead suggestion of that the acrosome is multifunctional, not laid restraint on single function at fertilization (*See* Artificial fertilization). In addition, it is worth noting that the proximal centriole is directly connected to fractional fragments of the nuclear envelope.

All features of *Urechis* sperm are different from the known in echinoderm sperm (Sato & Osanai, 1986; Hodgson & Bernard, 1986), but can be compared to those of mollusks or Annelids (Franzen, 1983: Russell & Peterson, 1983: Tilney, 1978: Tilney & Inoue, 1985). However, the shape and unique characteristics of *Urechis* sperm biased from those mollusks or Annelids are summarized as follows. The round disk-shaped nuclei with frontal concavity of *Urechis* sperm differ from streamlined or spheroid sperm heads of other marine animals. The sperm heads are arranged angularly to the flagella axis, while the heads of other marine invertebrates are arranged in a linear manner. The structural joining of the three mitochondria to establish a circular band or ring-shaped mitochondrion differs from the arbitrary or spiral arrangement of disjoined mitochondria covered with outer sheath. This type of joining in other marine invertebrates has not been previously described in the literature. Precursors of the acrosomal process (G-actin) are stored in the portion of the perinuclear theca in *Urechis* rather than the acrosome as is the case for other types of sperm.

Summary

It is known that *U. unicinctus* worm performs diffuse spermatogenesis. Male gametes are developed and differentiated in tissue level of coelomic fluid. But there is evidence of localized spermatogenesis, male gametes are developed in organ level of male storage sac. Sperm ball found in the male storage sac is the distinct unit of cells, consisted with developing male gametes and somatic cell. During breeding season, the diffuse spermatogenesis continues parallel with the localized spermatogenesis.

Diffuse spermatogeneis of *Urechis* is asynchronous, various developmental stages are prepared from the coelomic fluid. Primordial male germ cells are appeared as aggregates floating in the coelomic fluid. The germ cells are dissociated from aggregate at stage of spermatocytes which subsequently enter spermiogenesis constructing syncytium, spermatid cluster. Membrane fusion between spermatocyte is observed during formation of spermatid cluster. In progress of spermiogenesis, spermatid cluster develops central cytoplasmic mass and spermatids arranged along marginal sites of the cytoplasmic mass. Cytoplasmic connections join the syncytial cytoplasmic mass to number of the marginal spermatids. In advance of developments, spermatid cytoplasm flows into the cytoplasmic mass which increases greatly in amount, while cytoplasm of spermatids decrease continuously. For a while, spermatid nucleus is condensted and at final stage of developments, mature sperm detached from the spermatid cluster leaving cytoplasmic mass behind. The remained cytoplasmic mass is equivalent to residual body appeared in vertebrate spermatogenesis. The mature sperm seem stored in male storage sac. The process of sperm storage is not known.

No somatic cell is observed in the aggregate of primordial male germ cells and spermatid cluster. For the reason of coelomocyte occupies coelomic cavity in common with male germ cells, it is suggested that the coelomocytes support developments of the male sex cells. It is observed that structural differentiation and secretory activity of the coelomocyte coincides with cyclic development of the male sex cells.

Fine structure of mature sperm is differed from echinoderm, but compared with mollusk and Annelids. However, round and concave disk-shaped nucleus, massive acrosome and ring shaped mitochondria arranged perpendicular to axis of sperm head are unique features in *Urechis* sperm. The bipartite gradients of acrosomal materials around the central hollow are characteristic among marine invertebrates. Outer acrosomal materials functioned as adhesive in sperm-oocytes reaction and dispersed lately into vitelline layer, inner acrosomal materials is related with formation of acrosomal sheath. G-actin is stored in perinuclear theca continuous with central hollow of acrosome.

REFERENCES

Atkinson KH, KH Byram: The structure of the ovarian ball and ooenesis in *Moniliformis dubius* (Acanthocephala). J Morphol, 148: 391-426, 1976.

Awachie JB: The ecology of *Echinorhynchus truttae* Schrank, 1788 (Acanthocephala) in trout stream in North Wales. Parasitology, 55: 747-762, 1965.

Awachie JB: The development and life history of *Echinorhynchus truttae* Schrank, 1788 (Acanthocephala). J Helminthol, 40: 11-32, 1966.

Bondi C, WM Farnesi: Electron microscope studies of spermatogenesis in *Branchiobdella pentodonta* Whitman (Annelida, Oligochaeta). J Morphol, 148: 65-87, 1976

Choe RS, KS Shin, CN Joo, DY Hwang: The differentiation of the female gonial cells in Echiuroid *Urechis unicinctus*.: A fine structural study. Korean J Electron Microscopy, 18: 77-91, 1988.

Cooper JA: The role of actin polymerization in cell motility. Annu Rev Phyiol, 53: 585-605, 1991.

Crompton DWT: An ecological approach to Acanthocephalan physiology. Cambridge Univ Press, 1970.

Crompton DWT, PJ Whitefield: Observations on the functional organization of the ovarial ball of Moniliformis and Polymorphous (Acanthocephala). Parasitology, 69: 429-443, 1974

Fawcett DW: Observation on cell differentiation and organelle continuity in spermatogenesis. In "The genetics of the spermatozoon" . R A Beatty, S Gluecksohn-Waelsch eds. Edinburgh, pp 37-68, 1972.

Fischer A: The structure of symplasmic early oocytes and their enveloping sheath cells in the Polychaete, *Platynereis dumerilli.* Cell Tissue Res, 160: 273-343, 1975.

Franzen A: Ultrastructural studies of spermatozoa in three bivalve species with note on evolution of elongated sperm nucleus in primitive spermatozoa. Gamete Res, 7(3): 199-214, 1983.

Gould-Somero M: Echiura. In Reproduction of Marine Invertebrates, eds A. C. Giese and J. S. Pearse, 3: 277-311, Academic Press, Inc., New York, 1975.

Gould-Somero M, L Hollander: Fine structural investigation of the insemination response in *Urechis caupo*. Dev Biol, 46: 358-369, 1975a.

Gould-Somero M, L Hollander: Oocyte differentiation in *Urechis caupo*(Echiura): A Fine structural study. J Morphol, 147: 475-505, 1975b.

Hodgson AN, RTF Bernard: Ultrastructure of the sperm and spermatogenesis of three species of *Mytilidae* (Mollusca, Bivalvia). Gamete Res, 15:123-135, 1986.

Hoffmann, NH, CP Magnum: The function of coelomic cell hemoglobin in the polychaete, *Glycera dibranchiate*. Comp Biochem Physiol, 36: 211-228, 1970.

Jamieson BGM: A comparison of spermatogenesis and spermatozoa ultrastructure in Megascolecid and Lambridid earthworms (Oligochaeta; Annelida). Aust J Zool, 26: 225-240, 1978.

Jamieson BGM: Ultrastructure of spermatogenesis in *Phreodrilus phreodrillidas* (Oligochaeta, Annelida). J Zool, 194: 393-408, 1981.

Jeong JW, RS Choe, KS Shin, CN Joo: The differentiation of the male gonial cells in the storage sac of Echiuroid, *Urechis unicintus*: A fine structural study. Korean J Electron Microscopy, 19: 70-88, 1989.

Kierszenbaum AL, E Rivkin, LL Tres: Acroplaxome, an F-actin-keratin containing plate, anchors the acrosome to the nucleus during shaping of the spermatid heads. Mol Biol Cell, 14: 4628-4640, 2003

Korschelt E, K, Heider: Lehrbuch der Vergleichenden Entwicklungsgeschlechte der Wirbellosen Tiere. Jena: Gustav Fischer, 1902.

Kramer JA, SA Krawetz: Nuclearmatrix interactions within the sperm genome. J Boil Chem, 271: 11619-11622, 1996

Kwon HJ, KS Shin, WJ Kim: Fine structural investigation of fertilization envelope and acrosomal reaction in *Urechis unicinctus*. Korean J Electron Microscopy, 30(1): 61-72, 2000.

Kwon HJ, JW Jeong, WJ Kim, KS Shin: Immunofluorescence and electron microscopic study on the artificial insemination and rotation-shift behaviors of the bipolar spindle fiber in *U. unicinctus* egg. Korean J Electron Microscopy, 33(2): 105-116, 2003.

MacGuiness MP, MD Griswold: Interactions between Sertoli cells and germ cells in the testis. Seminars in Dev Biol, 5: 61-66. 1994.

Mayer A: Die Fruchung bebst Eibildung, Reifung und Befruchtung des *Gigantorhynchus gigas*. Zool Jahrb Abt F Anat, 50: 9-218, 1928.

Mayer A: In "Bronn's Klassen und Ordnungen des Tier Reiches". 2te Abt, 2te B, Acanthocephala. Leipzig Akad Verl Ges, 1932-1933.

Paul M: The polyspermy block in eggs of *Urechis caupo*: evidence for a rapid block. Exp Cell Res, 90: 137-142, 1975.

Russell LD, RN Peterson: Development of the acrosome and alignment, elongation and entrenchment of spermatids in procarbazine treated rats. Tissue and Cell, 15: 615-626, 1983.

Sawada N, K Ochi: Studies on fertilization in eggs of echiuroid *Ikedosoma gogoshimens* (Ikeda). I. An outline of the fertilization and the development. Mem Ehime Univ Sec II Natur Sci Ser Biol, B4: 437-445, 1962.

Sawada N, K Ochi: Electron microscope studies on sperm differentiation in marine annelid worms. Part I: Sperm formation *in Ikedosoma gogoshimens*. Dev Growth Differ, 17(2): 77-87, 1975.

Sato M, K Osanai: Morphological identification of sperm receptor above egg microvilli in the Polychaete, *Neanthes japonica*. Dev Biol, 113: 263-269, 1986.

Shin KS: Contribution of the body fluid to egg envelope formation in *Echinorhynchus truttae* (Acanthocephala). J Electron Microscopy, 35(3): 227-236, 1986.

Shin KS: The fine structure of sperm ball and sperm of *Urechis unicinctus* and immunogold localization of α-tubulin. Korean J Electron Microscopy, 28(2): 193-205, 1998.

Shin KS, Hj Kim, WJ Kim: Localization of anti-actin-gold particles (10nm) labeled to nuclear actin of *Urechis* sperm and spermatid. Korean J Electron Microscopy, 30(4): 403- 412, 2000.

Shin KS, WJ Kim: Fine structural study of coelomic solitary spermatogenesis in *Urechis unicinctus*. Korean J Electron Microscopy, 32(2): 107-119, 2002.

Shin KS, DH Lee, TY Ko: Cleavage pattern of Urechis unicinctus eggs in vitro fertilization. Korean J Electron Microscopy, 34(1): 71-81, 2004.

Shin KS, Kwon HJ, Kim WJ: Attribution of cortical granules to formation of fertilization envelope and polyspermy block in *Urechis unicinctus.* Integr Biol, 9: 57-64, 2005.

Soyer-Gobillard MO, J Ausseil, ML Geraud: Nuclear and cutoplasmic actin in dinoflaggelates. Biol Cell, 87(1-2): 17-35, 1996.

Tilney LG: The polymerization of actin IV. Role of Ca^{++} and H^{+} in the assembly of actin and in membrane fusion in the acrosomal reaction of echinoderm sperm. J Cell Biol, 77: 536-550, 1978.

Tilney LG, S Inoue: Acrosomal reaction of the Thyone sperm . III. The relationship between actin assembly and water influx during the extension of the acrosomal process. J Cell Biol, 100: 1273-1283, 1985.

Tres LL, AL Kierszenbaum: Variability of rat spermatogenic cells in vitro is facilitated by their cocultures of male germ cells of rats in contact with Sertoli cells in vitro. J Reprod Fertil, 57: 325-330, 1983.

Chapter 05

Cytological localization of nuclear actin and tubulin of sperm and spermatids in *U. unicinctus*

Chapter 05
Cytological localization of nuclear actin and tubulin of sperm and spermatids in *U. unicinctus*

5-1 Introduction

Spermatogenesis in *U. unicinctus* follows two patterns as described in the previous section. Localized and diffuse spermatogensis progress in the male storage sac (MSS) and body fluid, respectively. Localized spermatogensis in the MSS has been partly analyzed (Jeong *et al.,* 1989; Shin, 1998) but not fully elucidated as a whole, the details about diffuse spermatogenesis pertaining to the generation of the male gametes are better known (Shin *et al.,* 2000; Shin & Kim, 2002; Shin *et al.*, 2005b).

During diffuse spermatogenesis, large numbers of single male gametes developed individually in the coelomic fluid. At the beginning of spermiogenesis, individual male gametes are associated structurally with one another and form a syncytial spermatid cluster. Within the spermatid cluster, sperm shaping and nuclear condensation take place while individual spermatid develop into mature sperm. Most of the spermatid volume is eliminated during the processes of spermiogenesis. During spermiogenesis in *Urechis*, excessive cytoplasm flows out from the spermatid into the central cytoplasmic mass of the spermatid cluster, which is equivalent to residual body of the higher animals (Shin & Kim, 2002). The large spheroid nuclei of the spermatocytes are condensed into compact and discoid shapes enclosed in the syncytium of the spermatids. Since no somatic cell is enclosed in the syncytium, the reason of the nuclear condensation seems to be related to the solitary spermatid itself. A question arises how nuclear shaping and condensation of the solitary spermatid progresses while enclosed in the spermatid cluster.

It is known that intrinsic and extrinsic factors are involved in nuclear condensation of sperm in some animal species (Hecht *et al.*, 1986; Oliva *et al.*, 1987; Kierszenbaum *et al.,* 2003). Extrinsic factors include microfilaments of the somatic accessory cells and perinuclear microfilaments of the spermatids (Cooper, 1991; Soyer-Gobillard *et al.*, 1996; Kramer & Krawetz, 1996: Kierszenbaum *et al.,* 2003). In *Urechis* sperm, extrinsic factors can be excluded as factors that induce nuclear condensation since no somatic cells are associated with spermatids or in the spermatid cluster. Perinuclear manchette microfilaments are located near the nuclear envelope, but the microfilaments are rudimentary and not organized into a cylinder shape of the manchettes (Shin, 1998).

In the literature, intrinsic factors of nuclear condensation are involved in histone protein substitution with protamines (Hecht *et al.*, 1986), hyperacethylation of histone proteins (Oliva *et al.*, 1987; Christen & Dixon, 1983), and depolymerization of nuclear actin microfilaments (Cooper, 1991; Soyer-Gobillard *et al.*, 1996; Kramer & Krawetz, 1996). Because of the ontogenetic solitariness of *Urechis* male gametes, attention has been mainly paid to the intrinsic phase changes of actin microfilaments. To observe phase changes of actin microfilaments, *Urechis* male gametes and mature sperm are treated with FITC-phalloidin, anti-actin-FITC conjugate, and anti-actin-immunogold particles (AAIPs; 10 nm) (Shin *et al*, 2000, 2005). The actin microfilament polymer reacts with FITC-phalloidin (Faulstich *et al.*, 1977; Estes, *et al.,* 1981) while anti-actin-FITC depolymerizes actin (Kramer & Krawetz, 1996); AAIPs are incorporated into actin monomer (Cooper, 1991; Soyer- Gobillard *et al.,* 1996). The fluorescent dyes stained developing spermatid nucleus consistently, while AAIPs are differentially incorporated with developing spermatid nuclei. The number of incorporated AAIPs is increased in parallel with the progressive developmental stages of the spermatids. The parallel changes of the AAIP numbers are thought to be a consequence of successive disassembly of the actin microfilaments during nuclei condensation.

5-2 Nuclear condensation and shaping during spermiogenesis

5-2-1. Nuclear condensation

During spermiogenesis, nuclear downsizing and shaping are apparent and can be seen at every stage with electron microscopy (EM) (Shin & Kim, 1998). Much of the cytoplasm of the spermatocyte and spermatid is eliminated, and nuclei of the male gametes dramatically condense. The apparent decrease in spermatids cytoplasmic is parallel to the increase in the amount of central cytoplasmic mass of the spermatid cluster. Downsizing of cytoplasm can be directly identified by decreased dimensions of the nuclei and indirectly by increased densities of the nuclear matrices due to condensation. Even though some sperm organelles persist in the spermatid cytoplasm, the rate and degree of the shaping and condensation are rather rapid. It seems that some factors are continuously and actively involved in the shaping and condensation processes.

The reasons for changes in the shape and size of spermatocytes and spermatids have not been investigated in *Urechis* species. In some sense, *Urechis* male gametes are excellent for studying these reasons due to the solitary character of the cells. The solitariness minimizes the extrinsic factors of somatic accessory cells that can influence cell shaping and condensation.

During spermiogenesis, the spherical nuclei (5 × 6 μm) of the spermatocytes or primary spermatids are transformed into small and disc-shaped nuclei of mature sperm (1 μm thick with a 2 μm diameter). This downsizing is clearly seen in the spermatids of early stage of spermatid cluster (ESSC), at which time the oval spermatid nucleus measured 3 × 4 μm in size and its volume reached 27% of the primary spermatid nucleus. In the intermediate stage of spermatid cluster (ISSC), the spermatid nucleus decreased in size and volume with indentation of the frontal concavity, and measured 2.5 × 3 μm in size. In the late stage of spermatid cluster (LSSC) stage, the spermatid nucleus are 1 × 2 μm in size, with a volume that is approximately 2% of that of the primary spermatid. The LSSC spermatids are mobile and have fully developed flagella.

The nuclear matrices are rearranged during the course of nuclear condensation. Rearrangement and condensation of the contour can be observed in the processes of spermatid development. The nuclear matrices of the primary spermatids are homogeneous with a weak electron density. Homogeneity is largely retained, but the density is increased in spermatids enclosed in early stage of spermatids (ESSC). At the intermediate stage of spermatid cluster (ISSC), homogeneously dense nuclear matrices of the spermatids greatly increase in density. The nuclear matrices are granulated with numerous extra spaces between the granulated nuclear matrices, and a concave indentation appears at the front of the nucleus. As the late stage of spermatid cluster (LSSC) stage advances, nuclear vacuoles appear among the granulated nuclear matrices. It is likely that the appearances of nuclear vacuoles are related to fusion of extra spaces between the granulated nuclear matrices. Fusion of the extra spaces affects the nuclear condensation in three ways: decreased numbers of nuclear vacuoles, restoration of nuclear matrix homogeneity, and decreased nuclear volume. Consequently, nuclear matrices have high densities, and the volumes of the LSSC spermatids approach the volume of mature sperm. ISSC spermatids condense as nuclear vacuoles arise from the fusion of the extra space. At the beginning of LSSC stage, the number and size of nuclear vacuoles are maximized;

Table 1. Nuclear condensation in developing spermatids enclosed in different spermatid clusters and mature sperm. The nuclear matrix is greatly increased in density while its size is apparently decreased.

stages / features	Spermatocyte	ESSC spermatids	ISSC spermatids	LSSC spermatids	Mature sperm
Nuclear shape	Round, spherical	Round, spherical	Round, concavity	Disc, concavity	Disc, concavity
Nuclear size	5 × 6 μm	3 × 4 μm	2.5 × 3 μm	1 × 2 μm	1 × 2 μm
Nuclear density	diffuse	dense	dense, granular	dense, granular. nuclear vacuoles	dense, few nuclear vacuole
Volume (%)	100%	27%	12%	2%	Less than 2%

these are minimized again at later in the LSSC stage. Granulated matrices and nuclear vacuoles are observed in the spermatids as intermediate features during nuclear condensation.

Nuclear shape greatly changes during the late LSSC stage. The nucleus changes from a spherical or ovoid to a disc-shaped structure with frontal concavity. The distributions of nuclear shaping and condensation according to stage are summarized in Table 1. Some kinds of vectorial forces are in involved in the processes of spermatid condensation and shaping. These are related to actin microfilaments in the nuclear matrices.

5-2-2. Staining with fluorescence dyes and anti-actin-immunogold particle incorporation

According to the literatures mentioned above, a number of factors are involved in the processes of nuclear condensation. Phase changes of the nuclear actin microfilaments influence nuclear shape and condensation. F-actins in the nuclear envelope and matrices depolymerize into G-actin, initiating condensation and shaping of the spermatids. With the depolymerization of F-actin, vectorial forces of the F-actin disappear. Consequently, condensation of the nuclear matrices and envelopes are accompanied by release of G-actin.

Releasing G-actin due to F-actin depolymeriztion in course of the nuclear condensation is partly proved using immunofuorescence dyes (Shin *et al.,* 2000, 2005). As described above, staining with FITC-phalloidin, anti-actin-FITC, and AAIPs (10 nm) can be used to differentially visualize F- and G-actin in samples of developing spermatids (Shin *et al.*, 2005a). Fluorescence reflected in the corresponding actin microfilaments can be analyzed with confocal laser scanning microscopy (CLSM), and the incorporated AAIPs are observed with EM. Different fluorescence agents produce differential reactions according to nuclear condensation and shaping of the spermatids. Nuclear matrices of spermatogenetic cells stained in FITC-phalloidin show weak reaction, but acrosomal materials react with concentrated fluorescence (Fig. 5-1). Anti-actin-FITC reacts positively with nuclear matrices, especially with nuclear envelope, but exhibits negative reaction with acrosomal materials (Fig. 5-2). It is sometimes necessary to view contour line of spermatogenetic cell including flagella in the fluorescence contrast. Double staining in anti-actin-FITC and anti-tubulin-FITC reveal usual fluorescence intensities on the patterned structures of the spermiogenetic cells including nuclear matrices, but acrosomal material is not distinct from separate fluorescence (Fig. 5-3). On the other hand, anti-tubulin conjugate show reliable fluorescence staining along the entire flagella, but nuclear matrices exhibits a weak reaction (Fig. 5-4). Staining characteristics of the fluorescence dyes infer that acrosomal materials are consisted with polymerized form of actin microfilaments, while much more amounts of G-actin are available in nuclear matrices, especially in nuclear envelope.

The available G-actin is visualized in treatment with anti-actin-immunogold particles. The developmental stages and degree of condensation can be expressed in term of the reacted intensities of the agents, especially in the number of incorporated AAIPs. ISSC and LSSC spermatids respond more effectively to AAIPs, which enable the qualification and quantification of available G-actin at each of the spermatid developmental stages. On

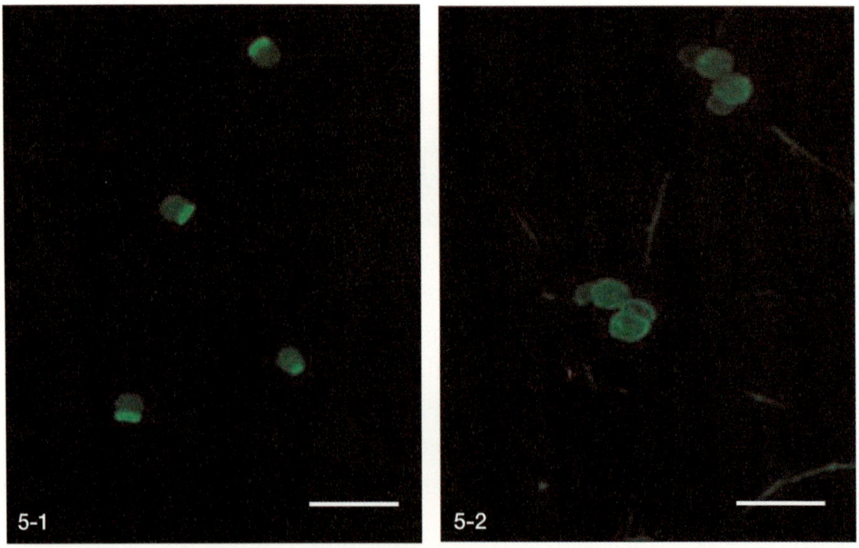

Fig. 5-1 Spermiogenetic cell reacted with FITC-phalloidin show positive reactions and bright fluorescence in acrosomal materials. Bar: 5 μm.

Fig. 5-2 Anti-actin-FITC react mainly in nuclear matices and nuclear envelope. Bar: 5 μm.

the other hand, spermatid acrosomes also react with the fluorescence dyes, and this staining can be associated with acrosomal function.

Specific Ag-Ab reaction of the AAIPs is defined as incorporation of two immunogold particles per μm^2 of the specimen in the EM section. Following treatment with anti-actin-AAIPs in *Urechis* spermatids, numerous AAIPs (~200 AAIPs) are incorporated into a gross area less than 4 μm^2 of the spermatid nuclei. Unspecific binding of AAIPs to the background is rarely observed. The number of incorporated AAIPs varies according to the developmental stage of the spermatids. On the other hand, because one gold particle has a standard capacity to bind fixed amounts of the Ag, the total number of AAIPs enables quantification of the available G-actin in the nuclear matrices. The quality of the actin microfilaments and quantities of G-actin varies during the different developmental stages of the spermatids. There are inclinations that number of incorporated AAIP in the nuclear matrices increased proportionally with developmental stages from ISSC to LSSC spermatid (Fig. 5-5).

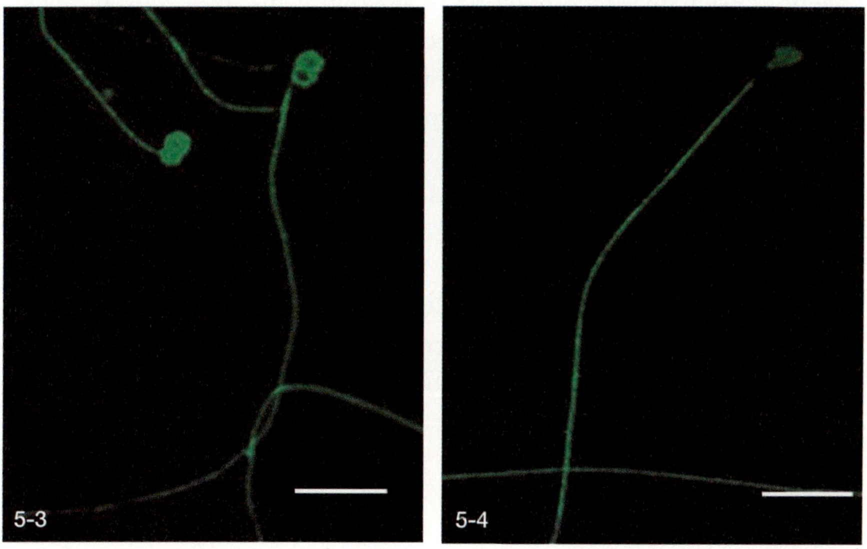

Fig. 5-3 Double staining in anti-actin-FITC and anti-tubulin-FITC reveal usual fluorescence intensities on the patterned structures of the spermiogenetic cells including nuclear matrices and flagella, but acrosomal material is not distinct from separate fluorescence. Bar: 5 μm.

Fig. 5-4 Anti-tubulin conjugate show reliable fluorescence staining along the entire flagella, but nuclear matrices exhibit a weak reaction. Bar: 5 μm.

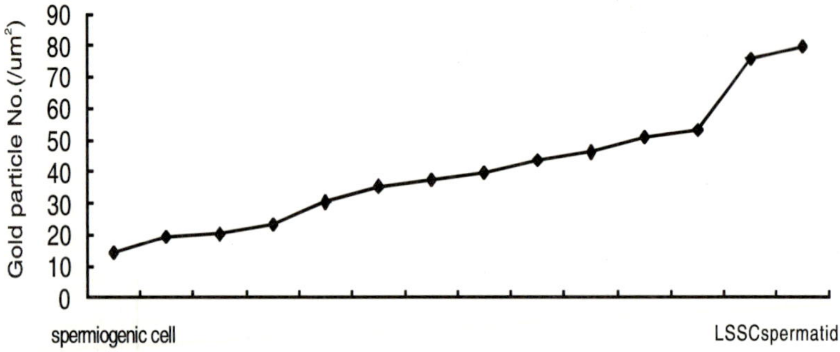

Fig. 5-5 During spermiogenesis, immunogold particles are incorporated into nuclear matrices, and there are inclination that number of incorporated gold particles is increased in advance of developemental stages from ISSC to LSSC spepermatids. The stage distribution is arbitrary ordered in referring to cellular differentiations, acrosome, mitochondria, condensation of nuclear matrices and compaction of nucleus.

Nuclear condensation and shaping can be expressed in terms of the quality and quantity of incorporated gold particles and the amount of available G-actin. Differential incorporation of AAIPs seems to be closely related with the appearance and disappearance of the extra spaces and vacuoles in the spermatids. The appearance of these nuclear structures indicates that nuclear condensation and shaping progress step by step or established grades at each developmental stage. The first step is the establishment of dense homogeneity in the spermatid nucleus at the ESSC stage. The second step is the appearance of the extra space between the granulated nuclear matrices at the ISSC stage (Fig. 5-6). The third and final step is the appearances of nuclear vacuoles at the LSSC stage (Fig. 5-7). The number of incorporated AAIPs is difficult to determine in the nuclei of ESSC spermatids, but peaks during the two other developmental stages. Over the ISSC to LSSC stages, the number of incorporated AAIPs is paralleled with the nuclear condensation, and is inversely proportional to the disappearance of extra space and nuclear vacuoles (Fig. 5-5). The extra space and nuclear vacuole seem fused each other at final stage of nuclear compaction, local concentration of the nuclear matrices and vacuolized region are distinctly observed within a nucleus. In this case, incorporated AAIPs are concentrated in the vacuolized region, less number of AAIP is found in the

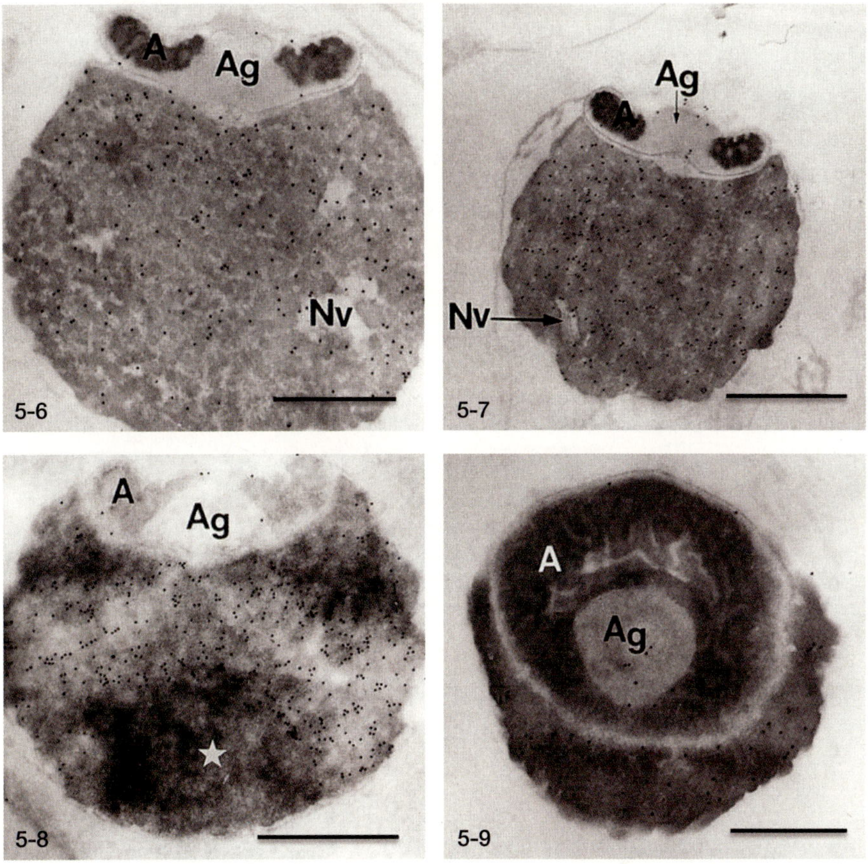

Fig. 5-6 Spermatid nucleus in developmental processes from ESSC to MSSC stage is in possess of extra space between granular nuclear matrices. Nuclear vacuoles appear in the nuclear matrices. Number of AAIPs are incorporated in the nucleus, especially at interface between vacuoles or extra space and nuclear matrices. Bar: 500 nm.

Fig. 5-7 Spermiogenetic cell at an advanced LSSC spermatid reveals more compact nucleus with a less extra space in the nuclear matrices. Much more number of AAIPs are incorporated in the nucleus. Bar: 500 nm.

Fig. 5-8 At final stage of nuclear condensation, AAIPs are distributed to the site of nuclear vacuole and extra spaces. It is likely that condensation and compact can be locally occurred, a less number of AAIPs are found in the condensed region of nuclear matrices (white asterisk). Bar: 500 nm.

Fig. 5-9 An oblique section through acrosome reveals that the AAIPs are mainly incorporated in the region of nucleus, a less number of AAIPs are to be seen in inner acrosome. Bar: 500 nm

condensed rerion of the nuclear matrices (Fig. 5-8). In addition, oblique section through the acrosome exhibits incorporation of AAIPs in nuclear

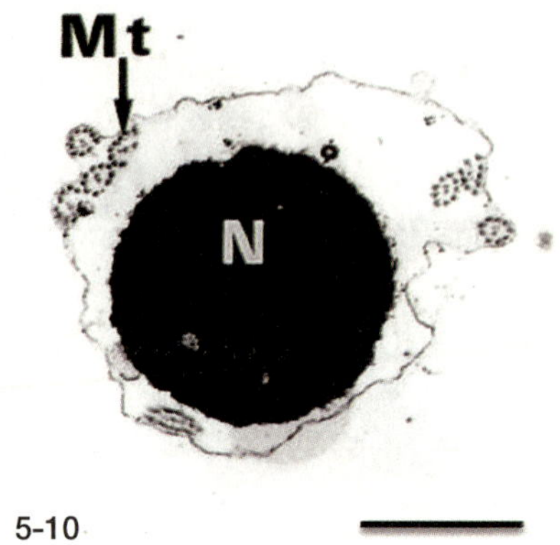

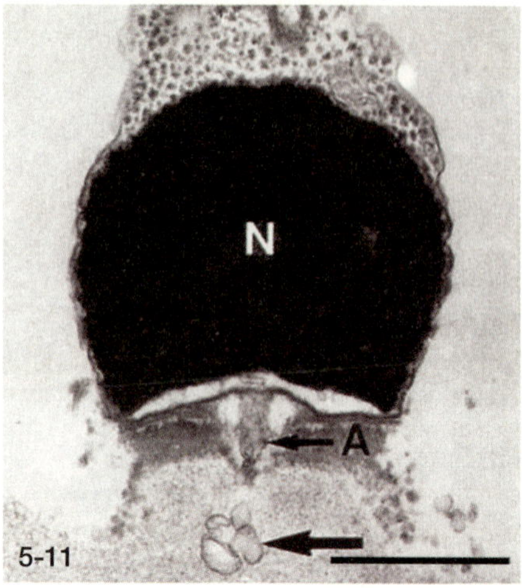

Fig. 5-10 Manchette microtubules (Mt), an extrinsic factor to condense nucleus (N) of spermatids are to be observed at around nucleus of the spermiogenetic cells. But the microtubules are not organized into cylinder form, suggestively out of function condensing nucleus. Bar: 500 nm.
Fig. 5-11 Condensed and compacted nucleus of mature sperm with less nuclear vacuoles is contacted and interacted to oocyte microvillus (block arrow). Bar: 500 nm.

matrices, but outer acrosomal materials shows no attached AAIP, while inner acrosomals exhibit a little number of attached AAIPs (Fig. 5-9).

Primary spermatid nuclei (5 × 6 μm) during the ESSC stage are reduced in size in ISSC nuclei (2.5 × 2 μm), which are ultimately condensed into the nuclei (1 × 2 μm) of the mature sperm. Because mature sperm nuclei are disc-shaped compact form with a little nuclear vacuole or extra space between nuclear matrices (Fig. 5-11), it seems that the contents of ISSC spermatid nuclei are condensed by vectorial forces into the discoid shape. In addition, AAIPs are incorporated consistently into the perinuclear theca. In the subacrosomal space and inner acrosomal materials, incorporation of AAIPs varies. Sometimes a small number of AAIPs are seen while no AAIPs are observed at other times. During fertilization, G-actin released from the perinucler theca passes through the subacrosomal components that are consumed during acrosomal assembly. Incorporation into the subacrosomal components is measured as released AAIPs from the perinuclear theca.

5-3 Extrinsic factors of nuclear condensation

In some sense, all kinds of cell nuclei undergo repeat cycles of nuclear condensation and expansion during the cell cycle. Consequently, different cell types contain nuclei of different shapes; this characteristic is often used as a criterion to identify cell types. It is in common in animal species, nuclear condensation and transformation of nuclear shape show species-specific configurations. The spermatocytes with large spherical nuclei are eventually condensed into tiny streamlined spermatids to enhance the hydrodynamic mobility in a fluid environment. During spermatogenesis, various extrinsic and intrinsic factors influence nuclear condensation and shaping. There are numerous extrinsic factors including manchette microtubules at the perinuclear space (McIntosh & Porter, 1967), ectoplasmic F-actin of Sertoli cell in mouse and rat (Kierszenbaum *et al.,* 2003), and somatic cells surrounding the male gamete (Cooper, 1991; Soyer-Gobillard *et al.,* 1996). The intrinsic factors include substitution of nuclear histone protein with protamine or with a histone-like basic protein (Hecht *et al.*, 1986), assembly and disassembly of actin microfilaments (Fath & Burgess, 1995), and hyperacetylation of histone proteins (Oliva *et al.*, 1987).

Since *Urechis* male gametes are free single cells floating in the body fluid, extrinsic factors of somatic cells are excluded as factors molding the nuclear shape of these cells. Even though male gametes are enclosed in the spermatid cluster during the later developmental stages, each spermatid maintains its solitariness. The spermatid cluster is in fact a syncytium consisted only of spermatids and no somatic cells are included in the syncytium enclosure. In addition, spermatid clusters do not seem contain accessory cells that support differentiation of male gametes and play a role in eliminating excessive cytoplasm.

The manchette cylinder, composed of perinuclear microtubules, is one of the primary extrinsic factors affecting sperm shape of various animal species (Witman *et al.*, 1978). Manchette microtubules are parallel arrays of microtubules in the perinuclear region oriented along the longitudinal axis of elongating spermatids. The cross-arms are coupled side by side with parallel arrays of interconnecting neighboring microtubules to form a cylinder

(Wilkinson *et al.*, 1974; Medina *et al.*, 1988). Assembly or disassembly of manchette microtubules affects spermatid nuclear condensation and shaping (Yoshida *et al.*, 1994).

The mature sperm of *U. unicinctus* appear as round discs unlike the spherical sperm heads of other marine species. All extrinsic factors involved in nuclear condensation and shaping described above seem to not affect *Urechis* sperm. On the other hand, manchette microtubules are to be observed in the perinuclear region of the spermatid nucleus. However, the microtubules do not seem to influence nuclear shaping in *Urechis*. Without the cross-arm, the microtubules are not organized into parallel arrays to form cylinders (Fig. 5-10). Instead, the perinuclear microtubules are located away from the nucleus and separated from each other (Shin *et al.*, 2005a).

In *Urechis* spermatids, the manchette microtubules are rudimentary and are not organized into a cylinder, nor do they help initiate nuclear shaping (Shin, 1998). Moreover, the nucleus of *Urechis* sperm is round and disc-shaped with frontal concavity. The round sperm head contains mitochondria without a fibrous sheath; this is remarkable in view of the evolutionary distances from vertebrates (Fawcett *et al.*, 1971, 1975; Longo & Anderson, 1968; Philips, 1980). The round shape of the sperm is not due to manchette microtubules. The manchette microtubules are inclined toward the posterior equatorial plate (*Avers*, McIntosh & Porter, 1967; *mouse, rat, rabbit, monkey*, Fouquet, 1992), while nuclear shaping progresses over the entire range of the nucleus inclusive of the frontal portion of the posterior equatorial plate. The sperm of other species are not streamlined, and manchette microtubules are not observed at the proper position in these cells (*Scorpion*, Philips, 1983; *Thyone,* Inoue & Tilney, 1982; *Crane fly,* Strauch, 1986, *Nematode*, Italiano *et al.*, 1999). The latter authors reported that nuclear shaping of the sperm head is based on nuclear condensation mediated by depolymerization of F-actin in the nuclear matrices.

5-4 AAIPs incorporation

Although *Urechis* male gametes exclude all of extrinsic factors involved in nuclear shaping and condensation, primary spermatids (5 × 6 μm) decrease in size to that of mature sperm (1 × 2 μm). Since actin microfilament is known as one of the intrinsic factors involved in nuclear shaping (Cooper, 1991; Soyer-Gobillard *et al.*, 1996; Kramer & Krawetz, 1996) and AAIPs localize in available G-actin in EM sections (De May, 1983; Slot & Geuze, 1989), immunofluorescence dyes and AAIPs are used to observe the available G-actin in developing male gametes. These dyes and AAIPs stain nuclear actin microfilaments and available G-actin in nuclear matrices (Shin *et al.*, 2000; Shin *et al.*, 2005a). In particular, AAIPs are differentially incorporated according to the developmental stages and stage distributions, and can be used to observe several stages of the condensation processes.

Overall, changes in AAIP incorporation proceed stepwise with nuclear condensation. The first step involves both of cytoplasm and nucleus. At the ESSC stage, excessive cytoplasm is eliminated from the primary spermatid while loosely-arranged or open-spaced nuclear matrices are rearranged into thicker or denser matrices. Incorporation of AAIPs is decreased at this stage, but increase as the nuclear matrices diminish in volume to 27% of the primary spermatid volume. The appearance of histone proteins with the protamine replacements and hyperacethylation (Hecht *et al.*, 1986; Oliva *et al.*, 1987) are possible in ESSC spermatids. The second step is granulation of the nuclear matrices. Granulation in the ISSC spermatids enables the nucleus to decrease in volume by 12% of the primary spermatid. The third step is vacuolization of the nuclear matrices which reduces the volume of ISSC spermatids to more than 2% of primary spermatids. Granulation precedes vacuolization and the extra spaces between the granular matrices merge into a relative low numbers of vacuoles so that the entire surface areas of the extra spaces are consequently reduced. The forth step is disappearances of the vacuoles in during the conversion of LSSC spermatids to mature sperm, resulting in a volume about 2% of that of primary spermatid. Even though one or more nuclear vacuoles always remain in the mature sperm, larger numbers of nuclear vacuoles disappear from LSSC spermatids until detachment from the spermatid cluster. The nuclear matrices have a higher density with mature sperm have a greatly reduced volume compared to primary spermatids.

As seen above, the maximum number of incorporated AAIPs is observed during the transition from the ISSC to the LSSC stages. When granules and vacuoles coexist in the nuclear matrices, the number of incorporated AAIPs is greatly increased and maximizes during granulation of the nuclear matrices. Size reduction of the spermatid nucleus parallels the increased numbers of incorporated AAIPs. Since the nuclear condensation of *Urechis* spermatid is not affected by any extrinsic factors, it is reasonable to hypothesize that available G-actin incorporated with the AAIPs is derived from disassembled intranuclear actin microfilaments.

5-5 Quantification of available G-actin

The densities and number of the incorporated AAIPs enable estimation of G-actin quantity. This quantification is based on two factors. First, the available G-actin is represented as Ag-Ab-specific reaction in solitary spermatid. Second, as nuclear condensation proceeds, the number of merged AAIPs is not fixed but varies according to the developmental stages. The number of incorporated AAIPs peaks at the transition between the ISSC and LSSC stages. Thus, nuclear condensation can be expressed by number of incorporated AAIPs during the stages of the spermatid developments. Specific Ag-Ab reactions are defined as the incorporation of two immunogold particles per μm^2 in an EM section (Jahn *et al.*, 1984; Piperno & Fuller, 1985). An anti-mouse-IgG gold particle (AAIP) recognizes 4 ng of Ag when the section is treated with AAIPs at a concentration of 1.17×10^{12}/mL. More than 200 AAIPs can be counted in a sectioned gross area of less than 4 μm^2 (50 AAIPs/μm^2). Generally, 800 ng/μm^2 of G-actin are estimated to be available for AAIP incorporation on EM sections. The efficiency and degree of nuclear condensation seem higher than expected, and it is unlikely that a static nucleus in the conventional cell cycle releases a significant quantity of G-actin.

Incorporated AAIPs are not distributed evenly or uniformly within the entire nuclear matrices, but are often concentrated in localized portions of the nuclear matrices. As development proceeds to the LSSC stage, the nuclear matrices are not homogeneous but are granulated with a mixture of vacuoles, remains of the extra spaces, and partially homogeneous nuclear matrices. AAIPs localize mainly on the surfaces of nuclear vacuoles. At times, AAIPs are distributed along the entire diameter of the nucleus and form a band. These observations indicate partial or local condensation of the nucleus, and the local condensation repeats separately around the whole nucleus increase condensation rates and lead to the characteristic nuclear shape (Shin *et al.*, 2005a). Coordinated disappearance of the nuclear envelope is prerequisite for local nuclear condensation.

5-6 Nuclear condensation in coordination with the nuclear envelope

As mentioned above, primary spermatids eliminate 98% of their nuclear volume as they develop into the LSSC spermatid nucleus. Decreased volume of the *Urechis* spermatid nuclei are explained by the phase changes of F-actin and subsequent release of G-actin from the nuclear matrices. However, phase change of nuclear actin microfilaments is necessary not sufficient for altering the shape of the cell, condensing cellular contents, and changing the nuclear shape (Soyer-Gobillard *et al.*, 1996; Holly and Blumer, 1999). Diminishing of the nuclear envelope and nuclear matrices is necessary for nuclear shaping and condensation as well.

Disassembly and assembly of polarized actin microfilaments in the membrane components are reported to be the basis of cellular and nuclear shaping (Fath & Burgess, 1995) while membrane flow has been implicated in cellular motion (Lee *et al.*, 1993; Heath & Holifeld, 1991). The cell and nucleus are not significantly affected by changes in volume even though the phase changes of actin microfilaments in the membrane components initiate cellular motion. Directional cellular movement is initiated at local sites and confined to the sites, and does not incur forcible changes of cellular volume. This can be attributed to the solidity of actin microfilaments as shown by the retrograde flow of membrane components. The solidity of actin microfilaments was found to be 7.3×10^{-26} Nm^2 (Gittes *et al.*, 1993), and is not sufficient to maintain cell shape or initiate nuclear condensation (Adams *et al.,* 1990; Egile *et al.*, 1999; Mallavarapu & Mitchison, 1999).

Phase changes of actin microfilaments in each nuclear matrix and membrane component are necessary but not sufficient to initiate cellular or nuclear shaping and condensation. Coordinative synergistic effects associated with the availability of G-actin in the nuclear matrices and nuclear envelope components seem to be necessary to simultaneously induce condensation and shaping in male gametes. The synergistic effects are necessary for the membrane components of both the nuclear vacuoles and nuclear envelope. In addition, various additional factors are required to generate the functional synergy effects. A coordinated molecular network of microtubule and structural myosin is needed for cell shaping (Hoey & Gavin, 1992; Theriot & Mitchison, 1992) which involves a set of actin binding proteins (Verkovsky and Borisy, 1993).

Local concentrations of AAIPs indicating the surface of vacuoles is one of the sites at which actin microfilaments change into G-actin. Fusion of the extra spaces in the granulated nuclear seems to represent an intermediate stage of nuclear vacuole development, and *Urechis* spermatids are characterized by the disappearance of vacuoles from the nuclear matrices. This process is referred to as "nuclear vacuole-mediate condensation" and involves nuclear condensation and shaping during *Urechis* spermatogenesis along with enhanced condensation efficiency in the solitary spermatids. It is difficult to estimate the efficiency rate of nuclear condensation because it is not known how long condensation takes. However, it is possible that the speed of nuclear condensations increases with continuous local condensation.

5-7 Shape of the acrosome

Acrosomal shapes of *Urechis* sperm are partly described in the "Spermatogenesis" section. Experiments using fluorescent dyes reveal details of the separate constituents of the acrosome according to acrosomal function. Analysis of the constituents and shapes of *Urechis* acrosomes provide results that are different from previous descriptions of acrosomal function and behavior during fertilization (Kwon *et al.*, 2000; Shin *et al.*, 2005a). Even though *Urechis* are often classified into the Echinoderm family, the acrosomal shape of *Urechis* sperm differs from that of other Echinoderms (Sato & Osanai, 1986; Hodgson & Bernard, 1986). Additionally, fertilization experiments revealed that the function of the *Urechis* acrosome is significantly different. As described in the results, the shape of the acrosome is similar to that of Annelids and mollusks in terms of size (Franzen, 1983), differential gradations (Russel & Peterson, 1983), and the presence of a central acrosomal hollow (Tilney, 1978). These characteristics are primarily observed by staining with FITC-phalloidin and anti-actin-FITC followed by confocal laser scanning microscope as well as EM observations.

Under EM, *U. unicinctus* acrosomes are found to be discoid, with band-like gradations that can be divided into inner and outer constituents. The bands around the inner half of the axial acrosomal hollow are less dense, while the outer circumferential bands are much more electron-dense (Kwon, 2000). The inner and outer materials of the bipartite acrosome are stained differently by the fluorescence dyes. The inner materials are not reactive against FITC-phalloidin or anti-actin-FITC, and weakly incorporate AAIPs. The outer materials are intensely stained by FITC-phalloidin but react weakly against anti-actin FITC. The outer materials are classified according to the existence or inclusion of polymerized actin microfilaments. Anti-actin-FITC binds tightly to actin monomers, while FITC-phalloidin bind more tightly to actin polymers than actin monomers, and shifts the equilibrium between polymer and monomer filaments towards polymer filaments (Faulstich *et al.*, 1977; Estes *et al.*, 1981).

5-8 Function of the acrosome

In *Urechis*, it has been suggested that the acrosomal materials have to play role in adhesive binding (*U. caupo*: Gould-Somero, 1975; Gould-Somero & Holland, 1975). A similar observation was reported in sand dollars in which the microfilaments function in bindin, which facilitates sperm-egg adhesion during fertilization and mediates the fusion of phospholipid vesicles (Glabe & Vacquier, 1978; Inoue & Tilney, 1982). In *U. unicinctus*, the acrosome is not limited functioning only as simple binding material, but is adapted for multifunction purposes associated with bipartite materials (Kwon, 2000: Shin *et al.*, 2005b).

During fertilization in oocyte, a tiny acrosomal process of 0.5 μm in length is created through the axial hollow of the acrosome. Even if a spermatozoon approaches the oocyte at the right angle, the outer acrosome touches the oocyte and acrosomal process simultaneously because of the insufficient length of the acrosomal processes and concave curvature of the outer acrosome. The extension of the acrosome is not greater than the marginal site of the concave outer acrosome. Moreover, the binding of the acrosomal process seems rather fragile. The binding of the outer acrosome supports and strengthens the fragile acrosomal process binding since the concave curvature of the outer acrosome fits into the convex curvature of the oocyte surface, assuring a rather large contact area on the oocyte. Outer acrosomal materials cover the entire surface of the fertilization cone during the early periods of contact. Firm binding of the acrosome is maintained until the spermatozoon has penetrated the vitelline layer and the oolemma is fused with the membrane components of the acrosomal process.

Disappearance of the outer acrosomal material occurs after the spermatozoon has infiltrated the ooplasm. The outer acrosomal material is dispersed as fine granules or coarse particles on the vitelline layer of the oocyte following infiltration. At beginning of dispersion, the particles can be distinguished from the vitelline layer by their thick electron densities and filamentous contours. The dispersions are regarded as intermix of the acrosomal materials into the developing oocyte vitelline layer. Over time, the densities of the particles are reduced. The acrosomal materials become intermingled and then unified with the vitelline layer.

Further studies concerning the fate of acrosomal materials involved nude oocytes (Kwon, 2000). For these experiments, the vitelline layer of the oocyte was removed by treatment with low concentrations of EDTA. When the nude oocytes are fertilized, the acrosomal materials do not disperse nor are they consumed. Instead, they remain on the oolemma while the zygote enters the cleavage stage. As might be expected, nude zygotes fail to form a fertilization envelope and are prevented from entering the cleavage stage. Moreover, staining with Propidium showed that nude zygotes have serious polyspermy.

Unlike the outer acrosomal material, the inner acrosomal material seems to be dedicated to forming the acrosomal sheath as described in the previous section. On the other hand, a significant portion of the inner acrosomal materials remain on the vitelline layer after infiltration of the spermatozoon. Upon dispersion of the outer acrosomal materials, the inner acrosomal material diffuse outward from the oocyte surface leaving pale conical traces behind the fertilization cone. These traces tend to align with the track along which the fertilized spermatozoon approached the oocyte. Even though numerous sperm are crowded close to the nude oocyte, a small number of sperm have been observed in the conically-diffuse area of the inner acrosomal materials.

Based on these observations, the acrosome is thought to have the following functions. First, the outer acrosomal material may support the binding of the acrosomal process to provide a wide contact area on the vitelline layer that plays a role in sperm adhesion to the oocyte. Second, the outer acrosomal materials is thought to reinforce the vitelline layer when intermixed into the vitelline layer; this is the most powerful and important barrier to prevent polyspermy. The reinforced vitelline layer inhibits polyspermy during the most critical time periods before the cortical reaction. No exact data is available, but there is indication that the prevention of polyspermy is less efficient without the outer acrosomal materials. Third, the inner acrosomal materials are not restricted to the vitelline layer. Some evidence of eccentric diffusion and pale conical trace of the diffusion may be related to the exclusion of other sperm that approach through the similar pathway of the advanced spermatozoon. However, this does not explain the back diffusion of the inner acrosomal material. Finally, a portion of the inner acrosomal material is dispersed as fine particles along the longitudinal axis of the elongating acrosomal process and promotes the formation of the acrosomal sheath. A similar process is observed in humans of which acrosomal materials form the sheath of the elongating acrosomal processes (Flaherty *et al.*, 1988; Escalier, *et al.*, 1997).

The previous points of view have not been explored in the literature. Experiments using nude oocytes have shown that acrosomal materials participate in blocking polyspermy. The outer acrosomal materials promote soft binding of the spermatozoon by intermixing with the vitelline layer to form the most substantial barrier against polyspermy. It is suggested that dispersion of the inner acrosomal materials is related to exclusive competition for fertilization. A rational explanation would be that once the spermatozoon-oocyte contact occurs, the prevention of polyspermy is not the sole responsibility of the oocyte. Harmonized cooperation of the oocyte and spermatozoon may result in a synergistic effect to prevent polyspermy. The details of this process are not yet known, but the vitelline layer may act as a substrate that causes diffusion of the acrosomal materials for the purpose of preventing polyspermy.

5-9 The site of G-actin storage

It has been reported in sea urchin and various marine invertebrates that elongation of the acrosomal process is morphological evidence of an acrosomal reaction (Schakmann *et al;*, 1978; SeGall & Lennarz, 1979; Bibring *et al.*, 1984; Clapper 1984). The true nature of acrosomal reactions is centrifugal polymerization and self-assembly of the actomere due to depletion of stored G-actin in the acrosome (Fath and Burgess, 1994). Polymerization of G-actin depends on changes in Ca^{2+} concentration (Brundage *et al.*, 1991; Hahn *et al.,* 1992), pH, and a set of actin binding proteins (Verkovsky & Borisy, 1993). However, localization of G-actin is species-specific (Fouquet, 1990). G-actin has been reported to localize in the anterior nuclear fossa (*Purpuratus*: Tilney, 1978; Verkovsky & Borisy, 1993), perinuclear fossa (bull, rabbit, hamster: Flaherty *et al.*, 1988), subacrosomal space (mouse: Prigent & Dadoune, 1993), subacrosomal bulges (rabbit: Fouquet, 1989), and fibrous sheath (human: Flaherty *et al.*, 1988; Escalier, *et al.*, 1997).

In *U. unicinctus* sperm, incorporation of G-actin-specific AAIPs is observed in various subcellular structures. Incorporated AAIPs have been seen in the subacrosomal fossa, perinuclear fossa, inner acrosome, and perinuclear theca. However, consistent incorporation has also been observed in the central region of the perinuclear theca, which is regarded as the principal site of G-actin storage in *Urechis* sperm. AAIPs found in the other subcellualr structures are thought to be AAIPs discharged from the perinuclear theca due to the sharp sensitivity of the sperm. G-actin in *Urechis* sperm diffuses from the perinuclear theca, and passed through the perinuclear and subacrosomal fossa into the acrosomal hollow, where G-actin is used to form the actomere.

On EM sections, incorporated AAIPs in the acrosomal region are always under 4~6 in number. The number of incorporated AAIPs enables quantification of the stored G-actin because it is known that these particles are incorporated with a definite amount of Ag with adequate concentrations of particles per ml of solvent as described previously. In a practical sense, the quantity of G-actin as determined by the number of the incorporated AAIPs show decreasing inclination, by the reason of the preparation processes, especially etching procedure used in the experiments (Shin *et al.*, 2005a). In fact, the number of incorporated AAIPs is different for every experiment and depends greatly upon the duration of hydrogen peroxide treatment.

5-10 Immunogold particle localization of alpha-tubulin

Microtubules are responsible for constructing the cellular skeletal system, transporting ions and macromolecules, forming the filtering apparatus of filter feeder (Bradbury & Oliver, 1980: Hofmann-Muenz, 1991), and enabling ciliary and flagellar movements of unicellular organisms (Akowska *et al*., 1982; Hannes & Foissner, 1992). All microtubules are composed of alpha- and beta-tubulin, and have similar fundamental structures (Wehland *et al*., 1983; Raff *et al*., 1997). These similarities can be seen in the tubulin and dynein system of animal sperm flagella (Warner & Satir, 1974; Gibbons, 1983; Philips, 1983; Ogawa *et al.,* 1977). The similarities in molecular composition and assembly are evidence of phylogenic conservation of the alph- and beta-tubulin and microtubules (Hiller, 1993; LeDizet & Piperno, 1987).

It seems important to know details about *U. unicinctus* sperm to better understand *U. unicinctus* biology because no unified classification system exists to determine the phylogenesis of the worm. Classification of *U. unicinctus* has not been precisely defined. Currently, the worms are classified as either *Echinoderm*, *Annelida* or mollusk. In fact, the external shape of the worm and characteristics of their sperm and spermatogenesis resemble that of *Annelida* and *Mollusk*, while some of the internal organ, egg characteristics, and cleavage patterns are similar to those of *Echinoderm.* Characteristics of sperm including shape and behavior during fertilization, including the acrosomal reaction, have often been used to categorize animals and examine phylogenic relationships. There is no previous literature that describes sperm and spermatogenesis of this worm.

The general morphology of the worm sperm is described in spermatogenesis section. Therefore, in an attempt to observe the consistency and similarity of microtubules, the worm sperm was treated with monoclonal anti-mouse tubulin antibody. Primary Abs binding was visualized by treatment with gold-conjugated goat anti-mouse IgG (AAIPs; Ris, 1985; Yang *et al*., 1997). The incorporated AAIPs were observed on the flagella tubulin, manchette microtubules, perinuclear theca, and centrioles. Flagella microtubules are the primary site of AAIP incorporation, indicating that these have higher affinity with the Abs. Presence of the gold particles are evidence of specific reactions between the flagella tubulin and monoclonal anti-mouse-tubulin.

These results suggest a phylogenic relationship between marine tideland animals and vertebrates. The similar associations have been reported between sea urchin and mouse (Weber, 1975; Harris & Rubin, 1987), and between cultured horse kidney cell and microtubules of sea urchin oocytes (Gorbsky & Borisy, 1985). Sequence conservation can be deduced by treatment with Abs that recognize the same Ag sites because the monoclonal Abs recognize a specific tyrosine residue (Gly)-$(Glu)_3$ -$(Gly)_2$ -Tyr) in the C-terminus. It can be inferred that the microtubules and axoneme of *U. unicincus* sperm contain the tyrosine residue in the C-terminal that is recognized by these Abs (Hiller, 1993; LeDizet & Piperno, 1978).

A quantify estimates roughly the amount of the Ag of the sperm flagella. The second Ab recognizes 4 ng of Ag when the gold particle concentration is 1.17×10^{12}/mL (Jahn *et al.*, 1984; Piperno & Fuller, 1985). For the qualitative analysis, however, the concentration of the gold particles was set at 5.85×10^{10}/mL and produced acceptable results. Various experimental factors influence the number of incorporated gold particles, especially the process and time for sample etching, and the number of incorporated AAIPs tends to decrease.

Microtubules of manchette and flagella are commonly found in animal sperm (Witman *et al.*, 1978). However, manchette microtubules are located around the nucleus. Rather than being organized into a cylinder shape, they are scattered at a distance from the nucleus without associated parallel array by cross connections. The sperm of *U. unicinctus* are different from the sperm of echinoderms. For example, the sperm head of *Urechis* is not spheroid, but are discoid in shape and include manchette microtubules (Fawcett *et al.,* 1971; Fawcett, 1975). Three mitochondria are fused to form a circular band located in cytoplasm of the posterior nuclear fossa that differs from the spiral arrangements of mitochondria in the fibrous sheath of advanced sperm (Longo & Anderson, 1968)

On the other hand, the sperm of *U. unicinctus* are remarkably similar in shape with Annelids (Sato & Osanai, 1986) and mollusk sperm (Hodgson & Bernard, 1986). These type of sperm are similar in shape, size (2 μm × 500 nm), mass (Franzen, 1983), and gradient concentrations of the acrosomal materials (Russel & Peterson, 1983). In addition, the central hollow (acrosomal space) constructed of polymerize G-actin (Tilney, 1978), fraction of the nuclear envelope extending into the proximal centriole, and direct connection of the distal centriole to the flagella are different from the sperm of echinoderm (Tilney, 1978), but are similar to that of mollusks (Hodgson & Bernard, 1986).

Summary

In the most animal species, large oval nucleus of spermatocyte is eventually condensed into tiny streamlined nucleus of spermatids to enhance the hydrodynamic mobility in a fluid environment. It is known intrinsic and extrinsic factors are involved in condensing and shaping of spermatocyte nucleus during the spermiogenesis. The extrinsic factors come from out of the generative cells, in the most cases from support cells associated together with the generative cells. Perinuclear microtubules revealed in anti-tubulin treatments are found at around the nucleus, but not organized into manchette, exclude from the extrinsic factors in *U. unicinctus*. Furthermore, *Urechis* sperm head is not streamlined in form, but round disk-shaped with frontal concavity.

Because of the ontogenetic solitariness of *Urechis* male gametes, attention has been paid to phase changes of intrinsic actin microfilaments. *Urechis* male gametes and mature sperm are treated with FITC-phalloidin, anti-actin-FITC conjugate, and anti-actin-immunogold particles (AAIPs; 10 nm). The nuclear matrixes exhibit weak but consistent reactions to the fluorescence dyes. The weak consistencies of the stainings reveal uniform existence of actin monomer and polymer in the nuclear matrixes. The proportional increase of the incorporated AAIPs according to progression of spermiogenesis is interpreted as consequence of depolymerization of actin microfilaments, which cause the nuclear condensation in association with compaction of nuclear envelope.

Fine structural changes of nuclear matrixes are in correspondence to the pattern of increased incorporation of AAIPs. During the continuation of nuclear condensation, density increase, granulation of nuclear matrixes with extra space between granules, vacuolization by fusion of the extra spaces, compact packing with disappearances of the vacuoles are serially observed in the nuclear matrixes. It is observed that the AAIP incorporation is closely related with appearances of vacuoles, the extra space between granules and nuclear vacuoles. There are inclination that much more number of AAIPs is incorporated at disappearance of the extra spaces and vacuoles. This process is referred to as "nuclear vacuole-mediate condensation". It is suggested that the interface of vacuole is the site of AAIP incorporation. On the other hand, FITC-phalloidin stained outer acrosome intensively and small number of AAIPs is incorporated inner acrosome. The staining characteristics of FITC-phalloidin suggest that the outer materials of acrosome are regarded containing much amount of polymerized actin microfilaments playing adhesive in sperm oocyte interaction.

REFERENCES

Adams AE, DI Johnson, RM Longnecker, BF Sloat, JR Pringle: CDC42 and CDC43, two addition genes involved in budding and the establishment of cell polarity in yeast *Saccharomyces cerevisiace*. J Cell Biol, 111: 131-142, 1990.

Akowska JJ, E.M Nelsen, Frankel J: Development of the ciliary pattern of the oral apparatus of *Tetrahymena thermophila.* J Protozool, 29(1): 366-382, 1982.

Bradbury PC, LS Olive: Fine structure of the feeding stage of a sorogenic ciliate, *Sorogena stoianovitchae* gen. n., sp. n. Protozool, 27(3): 267-277, 1980.

Brundage RA, KA Forgaty, RA Tuft: Calcium gradients underlying polymerization and chemotaxis of eosinophile. Science, 254: 703-705, 1991.

Bibring T, J. Baxandall, CC Harter: Sodium-dependent pH regulation in active sea urchin sperm. Dev Biol, 101: 425-435, 1984.

Clapper DL: Involvement of zinc in the regulation of pH, motility, and acrosome reactions in sea urchin sperm. J Cell Biol, 100: 1817-1824, 1984.

Christen ME, GH Dixon: Hyperacetylation of histone H4 correlates with the terminal, transcriptionally inactive stages of spermatogenesis in rainbow trout. Dev Biol, 93: 404-415, 1982.

Cooper JA: The role of actin polymerization in cell motility. Annu Rev Physiol, 53: 585-605, 1991.

De May JR: The preparation of immunoglobulin gold conjugated (IGS reagents) and their use as markers of light and electron microscopic immunocytochemisty, In:Cuello AC (ed), Immunohistochemistry, Wiley, New york, pp 347-372, 1983.

Egile C, TP Loisel, V Laurent, LR, Pantaloni, J Sansonetti, MF Cakier: Activation of the CDC42 effector NISP by the *Shigella flexnery* IcsA protein promotes actin nucleation by Arp2/3 complex and bacterial actin- based motility. J Cell Biol, 146: 1319-1332, 1999.

Escalier D, JM Gallo, J Schrevel: Immunochemical characterization of a human sperm fibrous sheath protein, its developmental expression pattern, and morphogenetic relationships with actin. J Histochem Cytochem, 45: 909-922, 1997.

Estes JE, LA Seldon, LC Gershman: Mechanism of actin of phalloidin on the polymerization of muscle actin. Biochemistry, 20: 708-712, 1981.

Fath KR, DR Burgess: Microvillus assembly. Not actin alone. Curr Biol, 5: 591-593, 1995.

Faulstich H, AJ Schafer, M Weckauf: The dissociation of the phalloidin-actin complex. Hoppe -Seylers Z Physiol Chem, 358: 181-184, 1977.

Fawcett DW, WA Anderson, DM Philips: Morphogenetic factors influencing the shape of the sperm head. Dev Biol 26: 220-251, 1971.

Fawcett DW: The mammalian spermatozoon. Dev Biol, 44: 394-436, 1975.
Flaherty SP, Winfrey VP, GE Olson: Localization of actin in human, bull, rabbit, and hamster sperm by immunoelectron microscopy. Anat Rec, 221: 599-610, 1988.
Fouquet JP: Immunogold distribution of actin during spermatogenesis in the normal rabbit and after experimental cryptorchidism. Gamete Res, 24(3): 281-290, 1989.
Fouquet JP: Immunoelectron microscopic distribution of actin in hamster spermadits and epidydymal, capacitated and acrosome reacted spermatozoa. Tissue Cell, 22(3): 291-300, 1990.
Fouquet JP: Species-spectific localization of actin in mammalian spermatozoa, fact of artifact? Microsc Res Tech, 20: 251-258, 1992.
Franze'n: Ultrastructural studies of spermatozoa in three bivalve species with notes on evolution of elongated sperm nucleus in primitive spermatozoa. Gamete Res, 7: 199-214, 1983.
Gibbons BH, IR Gibbons, B Baccetti: Structure and motility of the 9+0 flagellum of eel spermatozoa. J Submicrosc Cytol, 15: 15-20, 1983.
Gittes F, B Mickey, J Nettleton, J Howard: Flexual rigidity of microtubules and actin fimalments measured from thermal fluctuation in shape. J Cell Biol, 120: 923-934, 1993.
Glabe CG, VD Vacquier: Egg surface glycoprotein receptor for sea urchin sperm bindin. Proc Natl Acad Sci USA, 75: 881-885, 1978.
Gorbsky G, GG Borisy: Microtubule distribution in cultured cells and intact tissues: Improved immunolabelling resolution through the use of reversible embedment cytochemistry. Proc Natl Acad Sci USA, 82: 6889-6893, 1985.
Gould-Somero M, L Holland: Fine structural investigation of the insemination response in *Urechis caupo,* Dev Biol, 46: 358-369, 1975.
Gould-Somero M: Echiura. In Reproduction of Marine Invertebrates, eds A. C. Giese and J. S. Pearse, 3: 277-311, Academic Press, Inc., New York, 1975.
Hahn K, R DeBiasio, DL Taylor: Pattern of elevated free calcium and calmodulin activation in living cell. Science 359: 736-738, 1991.
Hannes A, W Foissner: Morphologie und Oekologie einiger Ciliaten (Protozoa: Ciliophora) aus dem Belebtschlamm. Arch Protistenkd, 141: 243-283, 1992.
Harris PJ, BP Rubin: Transition from Mitosis to interphase in sea urchin first division: Immunofluorescence studies of tubulin distribution in Methacrylate sections. J Histochem Cytochem, 35(3): 343-349, 1987.
Heath JP, BF Holifeld: Cell locomotion: New research tests on membrane and cytoskeletal flow. Cell Motil Cytoskeleton, 18: 245-257, 1991.
Hecht NB, PA Bower, SH Waters, PC Yelick, RJ Distel: Evidence for haploid expression of mouse testicular genes. Exp Cell Res, 164: 183-190, 1986.
Hiller SA: Ultrastructure of Prorodon (Ciliopora, Prostomatida). Somatic cortex and some implications concerning kinetid evolution in Prostomatid and Colpodid Cilisate. J Europ Microbiol, 40(4): 467-486, 1993.
Hodgson AN, RTF Bernard: Ultrasturcture of the sperm and spermatogenesis of three species of Mytilidae (Molluska, Bivalvia). Gamete Res, 15: 123-135, 1986.
Hoey JG, RH Gavin: Localization of actin in the *Tetrahymena* basal body-cage complex. J Cell Sci, 103: 629-641, 1992.
Hofmann-Muenz AH: The Oral apparatus of colpoda variabilis (*Ciliophora, Colpodidae).* II. Ultrastructure of the oral ciliature and its implications on ciliate phylogeny. Europ Protistol, 26: 288-302, 1991.
Holly SP, KJ Blumer: PAK-family kinases regulate cell and actin polarization throughout the cell cycle of *Saccharomyces cerevisiae.* J Cell Biol, 147: 845-856, 1999.
Inoue S, LG Tilney: Acrosomal reaction of thyone sperm. I. Changes in the sperm head visualized by high resolution video microscopy. J Cell Biol, 93: 812-819, 1982.
Italiano JE, M Stewart, TM Robert: Localized polymerization of the major aperm protein cy-

toskeleton correlates with the forward movement of the cell body in the ameboid movement of Nematode sperm. J Cell Biol, 146: 1087-1095, 1999.

Jahn R, W Schieber, P Greengard: A quantitative dot immunoblotting essay for proteins using nitrocellulose membrane filters. Proc Natl Acad Sci USA, 81: 1684-1687, 1984.

Jeong JW, RS Chae, KS Shin, CN Joo: The differentiation of the male gonial cell in the storage sac in Echiuroid, *Urechis unicinctus*: A fine structural study. Korean J Electron Microscopy, 19(1): 70-88, 1989.

Kierszenbaum AL, E Rivikin, LL Tres: Acroplaxome, an F-actin-keratin-constaining plate, anchors the acrosome to the nucleus during shaping of the spermatid head. Mol Biol Cell, 14: 4628-4640, 2003.

Kramer JA, SA Krawetz: Nuclear matrix interactions within the sperm genome. J Biol Chem, 271: 11619-11622, 1996.

Kwon HJ, KS Shin, WJ Kim: Fine structural investigation of fertilization envelope and acrosomal reaction in *Urechis unicinctus*. Korean J Electron Microscopy, 30(1): 61-72, 2000.

LeDizet M, G. Piperno: Identification of An acetylation site of chlamydomonas α-tubulin. Proc Natl Acad Sci USA, 84: 5720-5724, 1987.

Lee JA, A Ishihara, LA Theriot, K Jacobson: Principles of locomotion for simple shaped cell. Nature, 362: 167-171, 1993.

Longo FJ, E Anderson: The fine structure of pronuclear development and fusion in the sea urchin. J Cell Boil, 39: 339-368, 1968.

Mallavarapu A, T Michison: Regulated actin cytoskeleton assembly at filopodium tips controls their extension and retraction. J Cell Biol, 146: 1097-1106, 1999.

McIntosh JR, KR Poter: Microtubules in the spermatids of the domestic fowl. J Cell Biol 35: 153-173, 1967.

Medina A, FJ Moreno, G Garcia-Herdugo: Sperm tail differentiation in the nudibranch mollusc *Hypselodoris tricolor* (Gastropoda, Opisthobranchia). Gamete Res, 20: 223-232, 1988.

Ogawa K, T Mohri, H Mohri: Identification of dynein as the outer arms of sea urchin sperm axonemes. Proc Natl Acad Sci USA, 74: 5006-5010, 1977.

Olivia R, D Bazett-Jones, C Mezquita, GH Dixon: Factors affecting nucleosome disassembly by protamines in vitro. Histone hyperacetylation and chromatin structure, time dependence, and the size of the sperm nuclear preteins. J Biol Chem, 262: 17016-17025, 1987.

Philips DM: Observations on mammalian spermiogenesis using surface replicas. J Ultra-structure Res, 72: 103-111, 1980.

Philips DM: Analysis of sperm motility. J Submicroscop Cytol, 15: 29-35, 1983.

Piperno G, MT Fuller: Monoclonal Antibodies specific for an acetylated form of α-tubulin recognize the antigen in cilia and flagella from a variety of organism. J Cell Biol, 101: 2085-2094, 1985.

Piperno G, M LeDizet, XJ Chang: Microtubule containing acetylated α–tubulin in Mammalian cells in culture. J Cell Biol,104: 289-302, 1987.

Prigent Y, JP Dadoune: Immunoelectron microscopic distribution of actin in the spematids and epididymal spermatozoa of the mouse. Acta Anat, 147: 133-139, 1993.

Raff EC, JD Frankenthal, JA Hutchens, HD Hoyle, FD Tumer: Microtubule architecture specified by a β-tubulin. Science, 275: 70-73, 1997.

Ris H: The cytoplasmic filament system in Critical point-dried whole mounts and plastic-embedded sections. J Cell Biol, 100: 1474-1487, 1985.

Russel LD, RN Peterson: Development of the acrosome and alignment, elongation and entrenchment of spermatids in procarbazine-treated rats. Tissue Cell, 15: 615-626, 1983.

Sato M, K Osanai: Morphological identification of sperm receptors above egg microvilli in the polychaete, *Neanthes japonica*. Dev Biol, 113: 263-270, 1986.

Schakmann RW, WR Christen, BM Sapiro: The acrosome reaction of *Strongilocentrotus*

Purpuratus sperm: Ion requirements and movements. Dev Biol, 65: 483-495, 1978.

SeGall GK, WJ Lennarz: Chemical characterization of the components of the jelly coat from sea urchin egg responsible for induction of the acrosome reaction. Dev Biol, 71: 33-48, 1979.

Shin KS: The fine structure of the sperm ball and sperm of *Urechis unicinctus* and immunogold locatlization of α-tubulin. Korean J Electron Microscopy, 28: 193-205, 1998.

Shin KS, Hj Kim, WJ Kim: Localization of anti-actin-gold particles (10nm) labeled to nuclear actin of *Urechis* sperm and spermatid. Korean J Electron Microscopy, 30(4): 403- 412, 2000.

Shin KS, WJ Kim: Fine structural study of coelomic solitary spermatogenesis in *Urechis unicinctus*. Korean J Electron Microscopy, 32(2): 107-119, 2002.

Shin KS, HJ Kim, HJ Kwon, WJ Kim: Cytochemical localization of nuclear actin of sperm and spermatids in *Urechis unicinctus*. Integrative Bioscience, 9: 65-73, 2005a.

Shin KS, Kwon HJ, Kim WJ: Attribution of cortical granules to formation of fertilization envelope and polyspermy block in *Urechis unicinctus*. Integ Biosci, 9: 57-64, 2005b.

Soyer-Gobillard MO, J Ausseil, ML Geraud: Nuclear and cytoplasmic actin in dinoflagellates. Biol Cell, 87: 17-35, 1996.

Strauch AR: Biochemical analysis of actin in crane fly gonial cells: Evidence for actin in spermatocyte but not sperm. J Cell Biol, 86: 315-325, 1986.

Slot JW, HJ Geuze: Immunolabelling for electron microscopy. In "Techniques in immunocytochemistry" vol. 4 G. R. Bullock ed, pp 67-93, Elsvier Science, 1989.

Theriot JA, TJ Mitchison: The rate of actin based motility of intracellular *Listeria monogenes* equals the rate of actin polymerization. Nature, 357: 257-260, 1992.

Tilney LG: The polymerization of actin IV. The role of Ca^{++} and H^{+} in the assembly of actin and in membrane fusion in the acrosomal reaction of echinoderm sperm. J Cell Biol, 77: 536-550, 1978.

Verkovsky AB, GG Borisy: Non-sarcomeric mode of myosin II organization in the fibroblast lamellum. J Cell Biol, 123: 637-652, 1993.

Warner FD, P Satir: The structural basis of ciliary bend formation. Radial poke positional changes accompanying microtubule sliding. J Cell Biol, 63: 35-63, 1974.

Weber K: Specific visualization of tubulin Containing structures in tissue culture cells by immunofluorescence. Exp Cell Res, 95: 111-120, 1975.

Wehland J, J Mark, C Willingham, IV Sandoval: A rat monoclonal antibody reacting specifically with the tyrosinated form of alpha-tubulin. I Biochemistry characterization, effects on microtubule polymerization in vitro and microtubule polymerization in vivo. J Cell Biol, 97: 1467-1475, 1983.

Wilkinson RF, HP Stanley, JT Bowman: Genetic control of spermiogenesis in *Drosophila melanogaster*, the effects of abnormal cytoplasmic microtuble populations in mutant ms (3)10R and its colcemid-induced phenocopy. Ultrastruct Res, 48: 242-258, 1974.

Witman GB, J Plummer, G Sander: Chlamydomonas flagella mutants lacking radial spokes and central tubules. Structure, composition and function of specific axonemal components. J Cell Biol, 76: 729-747, 1978.

Yang ZH, GI Gallicano, QC Yu, E Fuchs: An unexpected localization of basonuclin in the centrosome, mitochondria and acrosome of developing spermatids. J Cell, 137(3): 657-669, 1997

Yoshida T, SO Ioshii, K Imanaka-Yoshida, K Izutsu: Association of cytoplasmic dynein with manchette microtubules and spermatid nuclear envelope during spermiogenesis in rats. J Cell Sci, 107: 625-633, 1994.

Chapter 06

Oocyte differentiation

Chapter 06
Oocyte differentiation

6-1 Introduction

Previous studies report that *Urechis* oogenesis differs in various points of view from oogenesis of other marine invertebrates including echinoderms, mollusks, and annelids. In *Urechis caupo*, the differences are the plural sites of oogenesis, timing of fertilization and cortical reaction, presence of cortical granules (CGs), and blocking mechanisms to polyspermy including formation of cytoplasmic indentation and nuclear invagination in the mature oocyte (Gould-Somero, 1975; Gould-Somero & Hollander, 1975b). Mature oocytes can be observed in both the coelomic cavity intermixed with coelomic fluid and female storage sacs (FSS). Some authors reported that oocytes stored in the FSS are the mature form capable for fertilization (Gould-Somero & Hollander, 1975a, 1975b). Other authors are of opinion that oocytes prepared from the FSS and coelomic cavity are premeiotic and young primary oocyte competent for fertilization (Shin *et al.*, 2005a; Shin *et al.,* 2005b).

Various studies have examined the appearances and presence of cortical granules (CGs) in *Urechis* oocyte. Some authors reported that CGs are found in oocytes, but do not take part in the cortical reaction and polyspermy blocking mechanism (Gould-Somero, 1975; Paul, 1975), while others described that the CGs and CG secretions are material bases of fertilization envelope preventing polyspermy (Shin *et al.*, 2005b). On the other hand, indentations of *U. caupo* oocytes are thought to be structural indications of oocyte maturity and fertilization capability. However, in *U. unicinctus* oocyte, indentation and nuclear invagination are subject to change due to the environmental conditions, and seem not exactly related with indication of oocyte maturity or the fertilization competency. These discrepancies seem associated with the unknown timing of meiosis which makes it difficult to evaluate the stages and stage distributions of *Urechis* oocytes.

In *U. unicinctus* oocytes, vitellogenesis and maturation precede meiosis, and premeiotic and postvitellogenic oocytes are competent for fertilization. Germinal vesicle is intact in the fertilization qualified oocyte and germinal vesicle break-down (GVBD) occurs after infiltration of spermatozoon into the cytoplasm of the oocyte. CGs are observed in the oocytes, the cortical reactions occur at fertilization and take part in polyspermy blocking processes (Shin *et al.*, 2005a, 2005b. *See* meiosis). CG secretions form a permanent fertilization envelope to block polyspermy and provide space for prospective embryonal development. However, there is a considerable delay between sperm-oocyte interaction and the occurrence of cortical reactions.

In most animal species, oocyte is defined as mature form of female gamete capable for fertilization arrested at a meiotic stage. Fertilization timing is shifted towards the premeiotic phase in *U. unicinctus*. Even maturation and vitellogenesis have been completed, the oocytes of *Urechis* are premeiotic gametes which are spawned and fertilized by a chance. The characteristics of *U. unicinctus* oogenesis are described in detail with special interest in stage distributions including meiotic and fertilization timing.

6-2 Developments of *U. unicinctus* oocyte

Mature oocytes prepared from female storage sacs (FSS) are capable of being fertilized, homogeneous in terms of differentiation and developmental stage. The oocyte developments continued in coelomic cavity are not synchronal; oocytes at various developmental stages are to be seen in preparations of coelomic fluid. The coelomic oocytes at the final developmental stage are similar to oocytes stored in the FSS in terms of size, structural differentiation, and fertilization competency. The mature coelomic oocytes are selectively stored in female storage sacs (FSS) (Gould-somero, 1975), even though the mechanisms of the selective storage are not yet known.

Developments of coelomic oocytes of *U. unicinctus* are classified into five groups including mature oocytes found in the FSS. Developmental stage I is oogonia enclosed in "oogonia cluster", which is an aggregate of a number of oogonia without associated somatic cell (Fig. 6-1). Stage II is characterized by the previtellogenic growth at which vitelline layer, microvilli and pinocytotic vesicles begin to appear on the oocyte surface (Fig. 6-2). During stage III, oocytes enter vitellogenic stage, yolk and cortical granules are to be observed in the ooplasm (Fig. 6-4). Germinal vesicle is

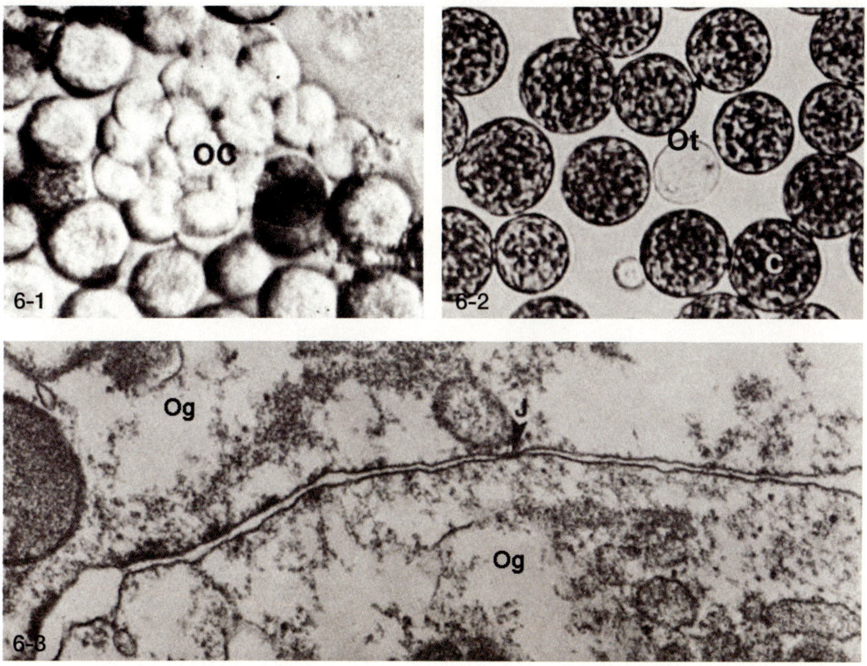

Fig. 6-1 Stage I living oogonia cluster (OC) prepared from coelomic fluid of *U. unicinctus*. The oogonia (Og) appear first as an aggregate embedded in coelomic fluid. Further investigations reveal no supportive somatic cell associated to the oogonia cluster. × 700.

Fig. 6-2 Stage II oocyte (Ot) is single cell dissociated from the oogonia cluster. The individual oocyte develops and differentiates solitarily floating in coelomic fluid. × 700.

Fig. 6-3 Junctional structures (J) are found between oogonia enclosed in oogonia cluster. But, the oogonial cluster seems not syncytium, each oogonium is separated each other, adheres simply to form the oogonial aggregate. × 50,000.

observed under interference microscope, appeared apparently in progresses of the oocyte developments. The stage III oocyte are vitellologenetic oocyte, various sized ranging from 70 μm to over 120 μm, even though similar contour line and external shape (Fig. 6-4, 6-5). The differences of the oocyte sizes come from accumulated plenty and amount of the cytoplasmic components, especially yolk granules. Consequently, the volume of the oocyte increases exponentially during the vitellogenic phase.

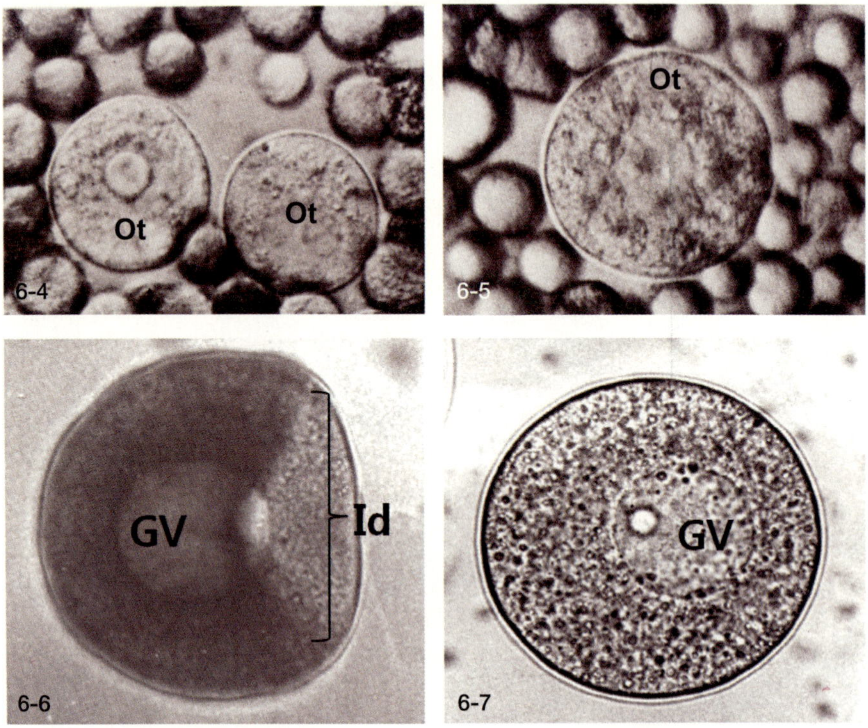

Fig. 6-4 The stage III oocyte (Ot) is measured 50 μm in diameter possessing germinal vesicle, distinguished from the coelomocyte in large size and transparency. × 700.

Fig. 6-5 Oocyte (Ot) of a late stage III is ranged 70 μm in diameter, become plentiful in cytoplasm and semi-transparency under light microscope. × 700.

Fig. 6-6 Stage IV oocyte is coelomic mature oocyte with cytoplasmic indentation (Id) and prominent germinal vesicle (GV). Some of the coelomic mature oocytes are competent for fertilization. Vitelline layer, the outer extracellular envelope is intact enclosing thin preivitelline space on oolemma. Oocytes prepared from the FSSs are similar in external shape with the coelomic mature oocyte. × 700.

Fig. 6-7 After spawning into brackish water, the indentation of both coelomic and stored oocyte is restored to round shape. Prominent nucleolus is to be seen in the germinal vesicle (GV). × 700.

Stage IV is postvitellogenic oocyte, characterized by appearance of prominent germinal vesicles including nucleolus, equivalent to the oocytes stored in the storage sacs in terms of maturity and fertilization capability as well as the external shape and size. Cytoplasmic indentation and nuclear invagination are observed in the stage IV oocyte as like stored oocyte in the FSS (Fig. 6-6). Coelomic mature oocyte is stored by unknown mechanism in the FSS, and these oocytes are spawned through sexual opening outside to meet a fertilization chance. The stage IV oocytes restore cytoplasmic indentation to round shape, when the oocyte is spawned into brackish water of lower salinity (Fig. 6-7).

6-2-1. Stage I oogonia

At the beginning of the breeding season around the end of January, oogonia are adhered together in clusters. Floating in the coelomic fluid, the oogonia cluster contains a number of oogonia (Fig. 6-1). The cluster is not a syncytium, but simple aggregate of oogonia without associated accessory or somatic cell. Each oogonium is semitransparent, measured 5-6 μm in diameter and distinguished from coelomocyte sharing the same coelomic cavity with the oogonia. Coelomocytes are 8 μm in diameter, opaque, and exist as single cells occupying the entire body cavity from time prior to the appearances of oogonia. Cell junctions are observed between oogonia cell membrane (Fig. 6-3). Marginal oogonia of the cluster are relatively round with wavy cell membranes, while oogonia at the center of the cluster are rather polygonal or rectangular in shape. Although dividing oogonia have not been observed, it is likely that the oogonia are in the proliferating phase. Within a short period of time, numerous oogonia clusters appear in the coelomic cavity.

It is hard to observe, when oogonium is dissociated from the oogonia cluster. Over time, it is likely that the oogonia separate from the cluster when they are about 8 μm~10 μm in size. The individual oogonia then enter into the previtellogenic growth phase.

6-2-2. Stage II oocytes in the previtellogenic growth phase

Stage II oocyte is individual previtellogenic oocyte (PrOs) floating in the coelomic fluid (Fig. 6-2). The oogonia are dissociated from oogonia cluster with size around 10 μm in diameter (Fig. 6-2). Round oogonia at the margin of a cluster are the ones first separated from the oogonia cluster. The separated PrOs in the coelomic cavity are distinct from coelomocytes. Although both PrOs and coelomocyte are similar in size and singleness, the PrOs are white in color and transparent, while the coelomocytes are shadowy and opaque. After separation, the PrOs continue to grow rapidly until reach 50 μm in diameter. Rapid growth of the PrOs is temporarily suspended when the PrOs reach 50 μm in size, and stand out conspicuously under LM.

A thickened preliminary vitelline layer (10-18 nm) is first developed on the PrOs. The actual thickness of the vitelline layer varies between sites because of the wavy surface of the cell membrane. A larger number of tiny microvilli appear being confined within the vitelline layer. More pinocytotic vesicles are observed, cytoplasmic contents are plentiful, and cell organelles are easily identified in the cytoplasm. Mitochondria are increased in number, scattered mainly in the cortex layer of the cytoplasm. Circular and concentric traits of the lamellate rER are ascertainable. Paraplasmic contents, such as dense inclusions, cytoplasmic aggregations, and lipid droplets, increase in number.

Nuclei of the PrOs reach about 20 μm in diameter. Nucleoplasm is characterized by a high electron density. Dense particles or scattered heterochromatin tend to remain aggregated and concentrated around the perinuclear space. The nuclear envelope is wavy with a frequently widened perinuclear space.

6-2-3. Stage III vitellogenic oocytes

Vitellogenic oocytes (VOs) enter into an exponential growth phase, sizes of the VOs are ranged from 70 μm to over 120 μm in diameter (Fig. 6-4, 6-5). Cytologically, the size and volume of the oocytes are greatly increased due to the increased yolk and cortical granules in number including various cell organelles. The cytoplasm attains substantial fullness and develops a different contour line compared to the previous stage.

The vitelline layer (0.4~1.5 μm thick) swells to uniform thickness and is enriched with indefinable fine substances. The microvilli increase in number, elongate, and project perpendicularly into the oolemma above the vitelline layer. Cortical granules are observed frequently in the cortex of the ooplasm, but their outline and size is unlike that of mature cortical granules. A few yolk granules appear in the cytoplasm, the VOs entered vitellogenesis. Since the VOs are single cells without associated somatic or accessory cell, yolk seems to be produced by autosynthesis involving substances absorbed from the coelomic fluid, which are the precursors of oocyte yolk. Pinocytotic vesicles increase in number and are scattered in the cortex of the ooplasm. Nuclear invagination is not observed, but a slight cytoplasmic indentation can to be seen in some of the VOs.

The nucleus is 25 μm in diameter. The nuclear matrices have substantial fullness, are homogeneous with electron dense materials, and contain a well-developed nucleolus (11 μm in diameter). Interchromosomal aggregates are scattered around the nucleoli and near the inside of nuclear envelope where more nuclear pores are developed.

Table 1. Developmental stages of *U. unicinctus* oocytes

Formation of cluster / Features	Cluster	Single cells								
		Microvilli	Vitelline layer	Pinocytotic vesicle	Yolk granules	CGs		Nucleolus	ID	S
Stage I	+	-	-	-	-	-	-	-	-	>10 (μm)
Stage II	-	+	+	+	-	-	-	-	-	50
Stage III	-	++	++	++	+	+	-	+	-	70
Stage IV	-	+++	+++	+++	++	++	++	++	+(+)	130
Oocyte in FSS	-	+++	+++	+++	+++	+++	+++	+++	+++	130~150

CGs: Cortical granules, FSS: Female storage sac
ID: indentation, S: size
+: Observed
++: Easily identified
++(+): Occasionally well-developed
+++: Well-developed

When the VOs are experimentally immersed in brackish water, simultaneous swellings all around the vitelline layer can be seen. The thickness of the swelled vitelline layer measured usually 5~11 μm thick, but slightly differ according to salinity of used brackish water. However, these swellings differ from the raising of fertilization envelope of the mature oocyte. Swelling of the VOs occurs simultaneously in the whole vitelline layer, while the vitelline layer of the fertilized oocyte is raises gradually, first from the portion around the fertilization cone then gradually all around the vitelline layer. The former is frequently confused with the raising of the fertilization cone. The developmental stages of the female gametes are ordered and summarized in Table 1.

6-2-4. Stage IV postvitellogenic oocytes

Postvitellogenic oocytes (PvOs) are mature cells that can be divided into two groups according to the appearing site of the oocytes: oocytes stored in the female storage sac (FSS) and coelomic oocytes. PvOs of *U. unicinctus* parepared from both sites are apparently premeiotic mature oocytes, equal in maturity and differentiation as well as fertilization competency. This equivalency has been demonstrated in artificial fertilization experiments. When coelomic mature oocytes are fertilized *in vitro*, some of them are cultivated into feeding larva.

Under LM (DIC), coelomic oocytes are homogeneous, pale yellow in color, and slightly transparent, thereby permitting an insight view of the intracellular components and nucleus. The stage IV is characterized by the frequent appearances of cortical granules aligned along the inside of the oolemma beneath (Fig. 6-9). The cortical granules (400~600 nm) are ovoid in shape, membrane-bound with electron dense contents, and sometimes closely approach the oolemma based on the presence of microvilli. Yolk granules (300 nm) observed in the living coelomic mature oocyte are bright yellow, ovoid in shape, and overwhelmingly outnumber of other inclusions scattered in the endoplasm of the coelomic oocytes (Fig. 6-8). The yolk granules are much larger and densely packed compared to the yolk granules of the previous stage. Lipid droplets are composed of unsaturated lipid, are irregular in form, and not numerous. The vacuoles (1.5-2 μm) are irregular in shape, and scattered rather loosely throughout the endoplasm, not as well-defined as yolk and cortical granules. The rER forms concentric lamellar circles that resemble the axonal myelin sheath of vertebrate nerve cell. Numerous microvilli extend outwards, and project through and above the surface of the vitelline layer (Fig. 6-9). When the coelomic oocytes are used in artificial fertilization experiments, it is often observed on EM that the sperm attach to the microvillous tip, which is believed to be the sperm receptor of the oocytes.

The stage IV coelomic oocytes contain a large nucleus (~50 μm) with a prominent nucleolus 10~14 μm in diameter. The nucleus is irregular in shape, due to the wavy and undulated nuclear envelope. The nucleolus is divided into the nucleolar core and cortex. Perichromatic particles are abundantly found around the nucleolus, and are particularly concentrated in close to the pars amorpha. Mitochondria are located in the deep portions

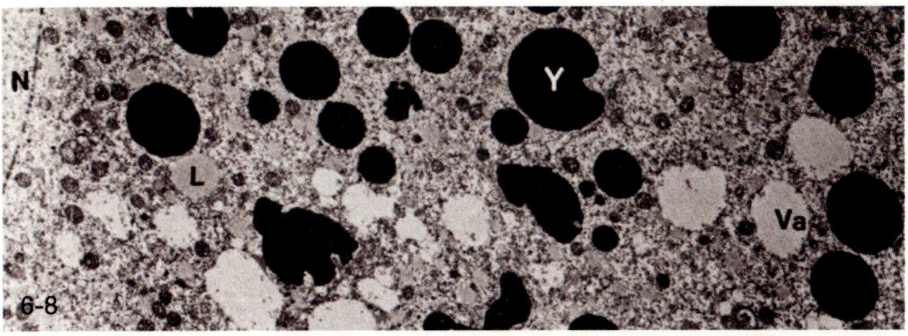

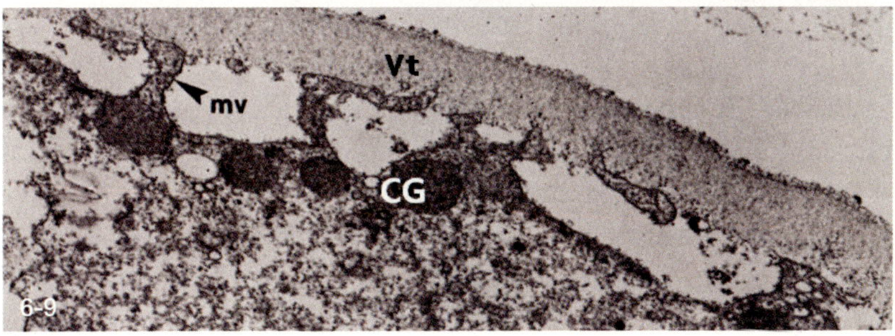

Fig. 6-8 Endoplasm of mature oocyte contains a number of yolk granules (Y), lipid droplets (L) and vacuoles (V). N: nucleus. × 6,000.

Fig. 6-9 Cortical granules (CG) are arranged in cortex region of mature oocyte, especially at the bases of microvilli (mv) which across perivitelline space to protrude beyond vitelline layer (Vt). × 20,000.

of the concave nuclear envelope.

The vitelline layers of the stage IV coelomic oocytes well-differentiated and can be divided into an outer thin surface coat and inner thicker vitelline layer. The inner vitelline layer is pale, wavy, and irregular in thickness (average thickness is 2.5 μm) because of the partial elevation of the oolemma at the base of the outward projections of the microvilli. The vitelline layer is rather uniform, 500 ~ 700 nm thick, and composed of depositions of fine and dense particles. The uniformity distinguished the vitelline layer from underlying perivitelline space, the boundary layer between the oolemma and the outmost vitelline layer. The perivitelline and vitelline layer serve as a site of cortical reaction and formation of fertilization envelope after occurrence of fertilization.

The cell membrane of the stage IV coelomic oocytes is uneven because of the membranous convexities at the microvillous bases. The oolemma and nuclear envelope are deeply invaginated, which is a unique characteristic of mature *Urechis* coelomic oocytes. Gould-Somero (1975) called the invagination of the oolemma an "indentation" to distinguish this feature from nuclear invagination. The depth of the indentation is about 39 μm in *U. unicinctus*, much deeper than the nuclear invagination (9 μm). In most cases, invagination and indentation turn the faces towards the same direction.

6-2-5. Oocytes stored in female storage sacs

Oocytes prepared from the FSS are 130~150 μm in diameter, slightly larger than coelomic mature oocyte. But, except for cell size, the differences between stored and coelomic oocytes are not easy to identify without practical measurement. In fresh samples, stored oocytes sometimes contain two or more indentations, while coelomic oocytes commonly exhibit one indentation. However, the indentations are flexible, and appear and disappear in response to different conditions. The roundness of the cells is restored as soon as it is immersed in brackish water.

As described above, stage IV coelomic oocytes are in the final developmental stage in the coelomic cavity. The coelomic mature oocytes are stored into two pairs of FSSs from which the stored oocytes are spawned to be fertilized. Fertilization efficiency of the stored oocyte is relatively high; approximately 80% of the stored oocytes are successfully fertilized in artificial fertilization experiments when conditions have been optimized including the correct ratios of oocytes and sperm, salinity, and temperature. On the other hand, some of coelomic oocytes are capable of being fertilized irrespective of lower fertilization rate. The rates of successful fertilization are based on the number of fertilized oocytes that develop into the cleavage stages. However, when coelomic and stored oocytes are mixed together in natural brackish water, and fertilized in addition of sperm suspension, it is difficult to distinguish the two types of cells from each another. The initial fertilizaztion responses of the coelomic oocytes are entirely identical with the stored oocytes. The oocytes prepared from the FSS contain intact germinal vesicle. The germinal vesicle can be clearly observed in both coelomic and stored mature oocyte. In addition, germinal vesicle break-down (GVBD) occurs after the fertilized spermatozoon has been entered the oocyte cytoplasm.

It is likely that the coelomic mature oocytes are stored in the FSS. To determine how oocytes are stored in the storage sac, Gould-Somero & Hollander (1975b) suggested the involvement of the thread-like structure projecting from the proximal site of the storage sac. In *U. caupo*, spiral curvatures of the thread structure are similar to the curvature of the indentation, and the compatible curvatures convey the coelomic mature oocytes into the storage sac. In *U. unicinctus,* the spiral thread is found at a similar site, but it is unlikely that the coelomic oocytes are transported into the storage sac along the spiral curvatures. Indentation of the *U. unicinctus* coelomic oocyte is a single invagination 39 μm deep, seems not compatible for the curvatures of the spiral thread.

SEM observations in *U. unicinctus* revealed a pore or opening at the distal tip of the storage sac (Fig. 6-10a, b). The sizes of the distal pore fluctuate between 24~100 μm, and coelomic oocytes crowded near the distal pore have been observed (Fig. 6-10c, d). However, the role of the pore in sorting coelomic oocytes from the mixtures of various developmental female gametes floating in the coelomic cavity is not clear. Stored oocytes are released through sexual pores, which are open to the outside and connect to the storage sac.

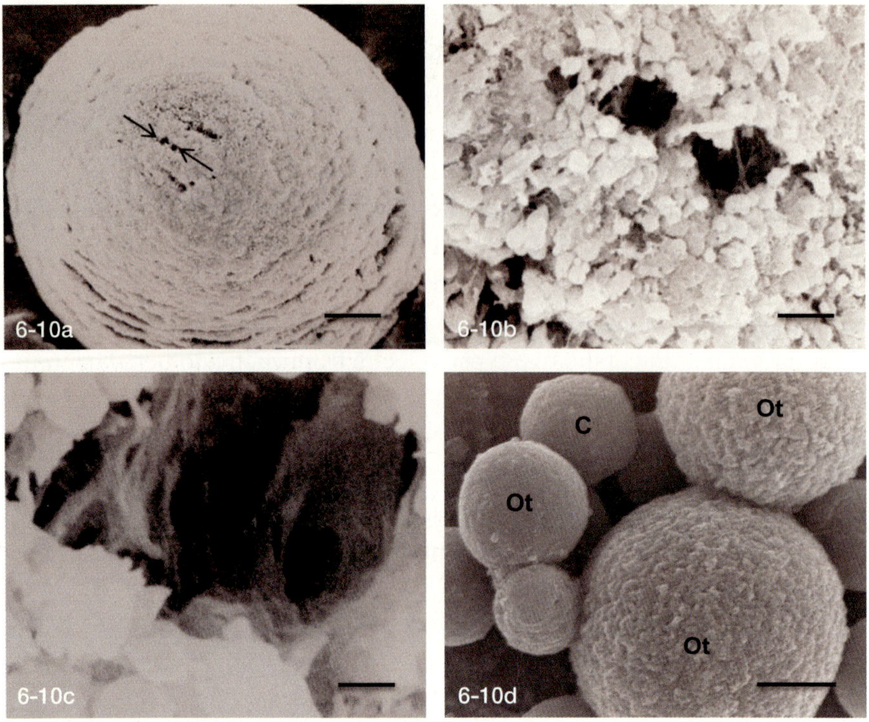

Fig. 6-10 (a) SEM observation reveals distal pores of various sizes (arrows) at the end tip of female storage sac. Bar: 100 μm. (b) An enlargement of the distal pore exhibits tissue organization in and around the distal pore. Bar: 10 μm. (c) The inside of the distal pore shows that the pore is not a simple pore, but organized compound tissue expecting a role in relation to female gametes. Bar: 1 μm. (d) Oocytes (Ot) and coelomocytes (C) are gathered and crowded around the distal pore and the crowd of cells has been observed during breeding season. Bar: 10 μm.

At the appearances of oogonia in the coelomic cavity, the storage sacs are not observed or exist as rudimentary organs hanged down the base of proboscis. When oogonia growth continues in the coelomic fluid and coelomic oocytes are produced, two pairs of FSSs begin to grow rapidly until they reach 5~6 cm in length and 5-8 mm in width. The FSSs are filled with oocytes from the proximal parts towards the distal parts. During the higher breeding season, extra partial expansions at the middle part of the FSSs are frequently observed. In advance of the breeding season, expansion of the FSSs are continues in parallel with maturation of the coelomic mature oocytes. These parallel events are the main reason why the stored oocytes are believed to be derived from the coelomic oocytes. However, the pathways or storage mechanisms of the mature coelomic oocytes have not been unequivocally elucidated.

6-3 Fertilization time of *Urechis* oocytes

As described above, the coelomic oocyte in the final differentiation stage is equivalent in fertilization capability to the stored oocyte in the FSSs which is eventually released for fertilization. Without considering meiotic timing in *U. caupo*, Gould-Somero (1975) referred to the stored oocyte a "full grown oocyte". The term "oocyte" represents the female gamete capable of fertilization. During oogenesis in most animal species, female gametes enter the meiotic phase after completion of vitellogenic maturation. Usually, meioses are stagnated at an interim stage from meiosis I and completion of meiosis II, and reactivation of oocyte development and completion of the meiosis is occurred after fertilization (*See* Meiosis). The female gamete under meiotic arrest is known as an "oocyte" which is competent for further development by chance of fertilization.

According to the literature, the earliest stage competent for fertilization is the young primary oocyte in some animal species. *U. unicinctus* oocytes are fertilized at the premeiotic stage with intact germinal vesicle. The germinal vesicle is not break-downed before occurrence of fertilization. Hence, the fertilization time of *Urechis* oocyte is shifted to pre-meiotic stage without GVBD and meiotic arrest.

However, coelomic and stored oocytes are called "oocyte" in the sense that both oocytes are ultimately spawned and successfully fertilized. Germinal vesicle break-down (GVBD) and meiosis begin 3~5 min after the fertilization. Two successive cycles of meiosis occur, producing three polar bodies in the fertilization envelope. The fertilized spermatozoon remains in the ooplasm for a while, until thc first meiosis is completed. The spermatozoon is activated at beginning of the second meiosis and fully activated at the end of the second meiosis initiating pronuclei fusion. These entire processes can be visualized using Propidium iodide and anti-tubulin-fluorescence staining methods followed by CLSM (*See* Meiosis).

6-4 Cytoplasmic indentation and nuclear invagination

In *U. caupo,* Gould-Somero & Hollander (1975; 1975b) first described cytoplasmic indentation and nuclear invagination of the *Urechis* oocytes. The indentation and invagination of oocytes are unique characteristics of *Urechis* oocytes, not commonly seen in other oocytes of animal species. The authors reported that the oocytes restore cytoplasmic indentation to round shape 10 min after fertilization, and recovery of the round shape is evidence of fertilization. Hence, the appearance of an indentation is visible evidence of oocyte maturity and fertilization capability. Indentation and its disappearance have been observed in fertilization experiments in both of coelomic and stored oocyte of *U. unicinctus*. In fact, indentations are found early in the most of postvitellogenic oocytes (> 100 μm) before they are stored in the storage sac. It is likely that the appearance of indentation is evidences of oocyte maturity. However, the coelomic oocytes are competent for fertilization, but not all of the coelomic oocytes is developed into cleavage stage. After fertilization, much many of coelomic oocytes are suspended developments showing abnormal or interrupted developments. Developmental failures have been frequently observed among fertilized oocytes prepared from the FSSs. However, ratio of the developmental failure or interruption is obviously decreased using the oocyte stored in FSSs.

In *U. unicinctus*, indentation formation and disappearance as well as the shape of the indentations are not consistent. Indentations of coelomic oocytes are different from each other. Some indentations are deeper and wider while others are shallow and narrow. Sometimes, a shallow indentation 70 μm in diameter can be seen in stage III vitellogenic oocytes. Typically, coelomic oocytes maintain one indentation while the stored oocytes exhibit deeper or more number of indentations.

Indentation flexibility has been demonstrated by immersing oocytes in brackish water. When oocytes are immersed in brackish water with the salt concentrations below 20‰, the indentations disappear and the cells gradually recover their roundness. When the oocytes are immersed in water with a salinity range of 23 ~ 27‰, the indentations are maintained, while deeper and more number of indentations arise in water with salinity above 28‰. Moreover, manipulation with a glass needle or higher temperatures can cause the indentations to disappear (Kwon *et al,* 2000).

Under natural conditions, the breeding seasons is commonly the mild rainy season during which a considerable amount of fresh water flows all over the worm habitat. Seawater turns into brackish water with salt concentration of 18~20‰. Even if the chance for fertilization is not available, the released oocytes recover their roundness in the natural brackish water. The appearance of indentations and recover of cell roundness are influenced by environmental changes. Thus, indentation cannot be a causal criterion to determine the maturity of an oocyte. Additionally, the restoration of an indentation is not a necessarily a consequence of the fertilization, but rather an adaptive structural device to compensate for flexible environmental conditions, especially in response to changes in salinity.

On the other hand, restoration of roundness in natural brackish water increases the chance for fertilization. Since the indentation is a deep concave part of the oolemma, sperm receptors on tips of numerous microvilli are concealed in the indentation (Kwon, 2003; Shin *et al*, 2005a). Once released into the brackish water of the worm habitat, the oocyte becomes rounded and the microvilli in the indentation are exposed uncovering the sperm receptors. The round oocytes with exposed sperm receptors increase the chances for fertilization compared to indented oocyte.

It has been noted that even when the concave oocyte is fertilized in seawater of higher salinity (23~27‰), the oocyte recovers its roundness similar to ones immersed in brackish water. In these cases, the necessary time for recovering the roundness varies in proportion to environmental salt concentrations. After fertilization, microvilli of the fertilized oocyte are flattened and retracted to accommodate the increased volume the fertilized oocyte and development of a thick fertilization envelope (15 μm). The raising of the vitelline layer or fertilization envelope at the fertilization cone is the first visible indication of fertilization under natural conditions. Once the oocytes recover their roundness, the fertilized oocytes remain round even when plunged into seawater of a higher salt concentration.

Summary

Urechis coelomic oocytes are classified into five developmental stages, oogonia cluster, perivitellogenetic and vitellogenic oocyte, including post-vitellogenetic coelomic oocyte and oocyte stored in the storage sacs. Individual oogonium is separated from oogonia cluster, enters perivitelline stage characterized by appearance of vitelline layer and pinocytotic vesicles. Yolk granules and cortical granules are accumulated in vitellogenic oocyte, and mature coelomic oocyte is qualified by abundances of yolk and cortical granules and appearance of cytoplasmic indentation. Except for slight large size, oocyte prepared from the storage sac is similar in external morphology to the coelomic mature oocyte. However, the stored oocytes are much more qualified for postfertilization developments. In addition, the formation of the indentation and invagination is conditional and flexible in accord to change of environmental factors, especially salinity of sea water.

Representative features of the mature oocyte can be referred as the massive germinal vesicle including substantial fullness. *Urechis* oocyte is fertilized at germinal vesicle stage.

REFERENCES

Gould-Somero M: Echiura. In Reproduction of Marine Invertebrates, eds A. C. Giese and J. S. Pearse, 3: 277-311, Academic Press, Inc., New York, 1975.

Gould-Somero M, L Hollander: Fine structural investigation of the insemination response in *Urechis caupo*. Dev Biol, 46: 358-369, 1975a.

Gould-Somero M, L Hollander: Oocyte differentiation in *Urechis caupo* (Echiura): A Fine structural study. J Morphol, 147: 475-505, 1975b.

Kwon HJ, KS Shin, WJ Kim: Fine structural investigation of fertilization envelope and acrosomal reaction in *Urechis unicinctus*. Korean J Electron Microscopy, 30(1): 61-72, 2000.

Kwon HJ, JW Jeong, WJ Kim, KS Shin: Immunofluorescence and electron microscopic study on the artificial insemination and rotation-shift behaviors of the bipolar spindle fiber in *U. unicinctus* egg. Korean J Electron Microscopy, 33(2): 105-116, 2003.

Paul M: The polyspermy block in eggs of *Urechis caupo*: evidence for a rapid block. *Exp Cell Res*, 90: 137-142, 1975.

Shin KS, HJ Kim, HJ Kwon, WJ Kim: Cytochemical localization of nuclear actin of sperm and spermatids in *Urechis unicinctus*. Integrative Biosciences, 9: 65-73, 2005a.

Shin KS, HJ Kwon, WJ Kim: Attribution to cortical granules to formation of fertilization envelope and polyspermy block in *Urechis unicinctus*. Integrative Biosciences, 9(2): 57-64, 2005b.

Chapter 07

Contribution of cortical granules to the formation of fertilization envelope and polyspermy block in *Urechis unicinctus*

Chapter **07**

Contribution of cortical granules to the formation of fertilization envelope and polyspermy block in *Urechis unicinctus*

7-1 Introduction

In many marine invertebrates, it is known the fertilization envelope (FE) plays important roles in processes of fertilization. One is the permanent block to polyspermy, and the other is the preparation of extracellular space for the fertilized oocyte proceeding prospective cleavage and further embryonic developments (Berg and Wessel, 1997). Cortical granules (CGs) are oocyte-specific secretory granules (Matese *et al.,* 1997) and consumption of CG secretions during FE formation and polyspermy block have been investigated mainly in marine invertebrate inducing sea urchin. During the cortical reaction, CGs are approached and located closer to the oolemma, fused eventually with the oolemma, and the contents of the CGs are expelled into the expanding perivitelline space, which develop at immediate future into permanent FE.

The cortical reaction is mediated by free Ca^{2+} and several membrane proteins, and the entire processes resemble exocytosis of somatic secretory cells (Conner *et al.*, 1997; Stricker, 1999). The temporary or fast block followed by the permanent block in expense of the CG secretions prevent basically the polyspermy (Fisher & Rebhun, 1983; Inoue & Tilney, 1982; Tilney & Inoue, 1985; Navare & Schatten, 1994). However, there are various cases in concern with presence and function of CGs in animal species. In echinoderms, mollusks, and some annelids, cortical reactions occur and the FE formation is based on the CG secretions (Vacquier, 1975), while

CGs in *C. elegns* and *D. melanogaster* have not yet been reported (Vacquier, 1981). On the other hand, although the CGs are appeared in processes of oocyte developments, its behavior and function are not regular, but variable and differ from species to species.

CGs are found in *Mytilus edulis,* but the CGs are discharged prior to fertilization when the oocytes are exposed to sea water, and the FE fails to block polyspermy (Humphreys, 1967). The flexibility in appearance of CGs and the contribution of CG secretions to the polyspermy block have been reported in *Urechis*. CGs are not found in *U. caupo* (Gould-Somero & Hollander, 1975b), and the FE fails to block polyspermy. On the other hand, CGs are found in *U, caupo*, but not associated in the FE formation, even though the FE is effective to prevent polyspermy (Paul, 1975). In this case, CGs maintain their original configuration until the fertilized eggs develop into embryos (Gould-Somero & Hollander, 1975b). It has been suggested that the mechanism of polyspermy block appears to be fundamentally different in *U. caupo* compared to other animals (Gould- Somero & Hollander, 1975a). The consumption of CG secretions for the formation of the FE and prevention of polyspermy varies from species to species (Barros & Yamaguchi, 1971).

Some of the characteristics of CG secretions in *U. caupo* differ from those in *U. unicinctus*. These differences are due to the delayed cortical reaction in *U. unicinctus* (Kwon *et al.,* 2000, 2003; Shin *et al.,* 2004). In *U. unicinctus,* CG secretions are not released soon after sperm-oocytes interaction. Prior to CG secretion, the perivitelline space reaches the full thickness and CGs are secreted into the thickened perivitelline space. CG secretions are not directly cause elevation of the FE, because CG secretions are delayed until the perivitelline space is completely thickened. The perivitelline space including the CG secretions differentiates into the final FE.

Until the cortical reaction occurs, the raised vitelline layer plays an important role in blocking polyseprmy. After the CG reaction, the FE serves as the permanent barrier against polyspermy and provides extracellular space devoted to the development of embryos. Delayed CG secretions in *Urechis* differ from observations made in the echinoderms, mollusks and annelids in which CG reactions occur immediately after the interaction between sperm and oocytes.

7-2 Sperm-oocyte interaction and cortical reaction

7-2-1. Sperm-oocyte interaction

Oocytes in the coelomic cavity are found to be intermixed with coelomocytes and coelomic fluid. Oocytes in the storage sacs are uniform in size (130-150 μm) and developmental stage, not demanded forcibly extra practical separation steps. Because of the homogeneous maturity and simple procedures for handling the oocytes, it is preferable to use oocytes isolated from the storage sacs. The oocytes are semitransparent with a pale yellow background and equipped with numerous microvilli that protrude above the vitelline layer. Prominent germinal vesicles (GV, 50 μm in diameter) and a nucleolus (3 μm in diameter) are observed under light microscopy (Fig. 7-1, insert).

The oocytes are artificially fertilized by mixing appropriate amounts of sperm into an oocyte suspension in natural sea water (NSW of 20‰ salinity, pH 8.1~8.2, at 15~17℃). NSW gathered from the ocean is in reality a type of natural brackish water because the higher breeding seasons correspond to the mild rainy season in the tideland where the worms live. Salinities of NSW vary during the rainy seasons, but are usually maintained between 18~22‰. Natural spawning and fertilization occur over a limited range of salinity and temperature.

Under the experimental conditions, indentations and invaginations are slowly and continuously disappeared. The oocytes recovered their round shape, and the microvilli along with the sperm receptors are exposed. The FE is formed within 15 min after the addition of sperm to the oocyte suspension. In addition, even though the oocytes are fertilized in sea water with a higher salinity, the indentations of the fertilized oocytes disappeared and the cells recovered their round shapes. Higher salinity sea water and temperature caused abnormal development of the fertilized oocytes while a lower temperature significantly delayed fertilization events and development.

Due to the translucency of the oocytes, initial sperm-oocyte binding is recognizable with confocal and interference microscopy. The visible events of fertilization include raising of the fertilization cone and vitelline layer, recovery of the round oocyte shape, retraction of microvilli into the raised vetelline layer, germinal vesicle break-down (GVBD), cortical reactions followed by FE formation, and two successive cycles of meiosis and cleavage. Except for GVBD and meiosis, other events of fertilization are closely related to the processes of FE formation and CG secretion.

7-2-2. Cortical reaction and formation of the fertilization envelope

The final FE is formed after 15 min of sperm addition. As soon as the sperm interacted with the oocytes, the raised vitelline layer spread out on the entire oolemma until the perivitelline space reached 15 μm thick. Next, the CGs are secreted into the thickened perivitelline space 5 min after sperm addition. The secreted CGs dispersed and formed five layers in the perivitelline space. Over time, materials in each of the five layers intermingled together to form a homogeneous monolayer exclusive of the outermost surface coat.

Formation of the final FE is completed within 15 min under the given conditions (NSW of 20‰ salinity, pH 8.1~8.2, at 15~17℃). For convenient description, the perivitelline space without CG secretions is called as initial FE (iFE) and differentiating FE containing CG secretions is termed as intermediate FE (itFE). In the final FE (fFE), the CG layers fuse with one another to form two layers; an outer thin surface coat and a thick homogeneous inner layer.

1) Formation of the iFE

For fertilization, oocytes are normally suspended in NSW prior to the addition of sperm. The cytoplasmic indentation disappears spontaneously and the oocytes recover their round shape, exposing most sperm receptors on the microvillous tips. As above, the cytoplasmic indentation does not recovered in seawater of higher salinity. However, the cytoplasmic indentation are recovered round shape, although oocytes are intermixed with sperm suspension and fertilized in seawater of higher salinity. The recovery of the indentation seems closely related to increased oocyte permeability.

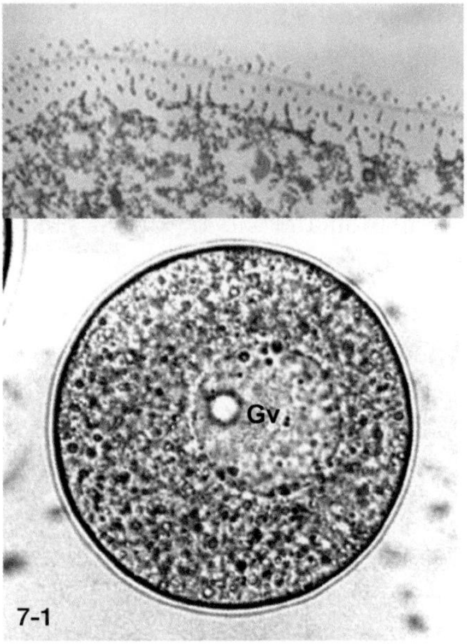

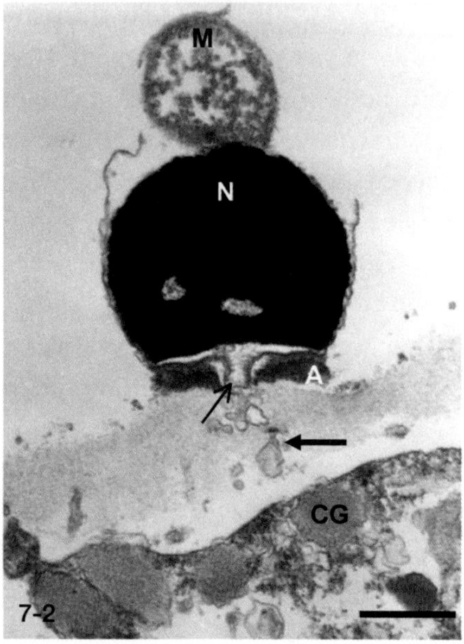

Fig. 7-1 The intact germinal vesicle (GV) at slightly eccentric position is an indication of the unfertilized oocyte. × 400. Insertion: numerous microvilli are projected onto the oolemma. Bar: 2 μm.

Fig. 7-2 Actomere (arrow) of the spermatozoon fused with the microvillous tip (block arrow), sperm receptor, while the massive acrosomal materials (A) adhered to wide range of the oocyte vitelline layer. Dense particles are disjoined from the acrosomal material dispersing on the vitelline layer. The cortical granules (CGs) are aligned closely oolemma beneath. Bar: 1 μm. M: Spermatozoon mitochondria, N: Spermatozoon nucleus.

The vitelline layer started to rise at the site of the fertilization cone and expanded horizontally over the entire surface of the fertilized oocyte. The expansion continued until the initial FE is 14~16 μm thick 5 min after the addition of sperm. When the outmost vitelline layer is 2~3 μm thick, most microvilli with sperm receptors are retracted and concealed in the perivitelline space (Fig. 7-2). Numerous sperm are still highly active and crowded around the oocyte until the thickness of the perivitelline space is increased. However, as the perivitelline space thickened over time, the sperm are inactivated and the raised iFE pushed the sperm from the oocyte. Given that the sperm have a short half-life, the iFE serves as a potent barrier against polyspermy during the most critical time period in which the sperm are still alive and active.

The term "iFE" is used to accentuate the potent ability of the FE to push surplus sperm away from the oocyte. This is demonstrated experimentally using nude oocytes that lack a vitelline layer. The iFE is 15 μm thick, but in a practical sense the perivitelline space could block polyspermy when it reached 2~3 μm thick. Staining with Propidium iodide followed by CLSM showed the withdrawal of numerous surplus sperm outside of the raised iFE. The distance between the oolemma and excessive sperm is increased proportionally until the iFE reached its full thickness (Fig. 7-3). The inside of the raised iFE produced a reflect image with radial arrays under confocal microscopy. Under EM, numerous retracted microvilli are observed (Fig. 7-4).

On the other hand, it is revealed that massive outer acrosomal material plays role in adhesive. During the course of EM observation, sections through the median plane angles of the sperm-oocyte interaction exhibited the entire frontal surface of the discoid acrosome (1 × 2 μm), which re-

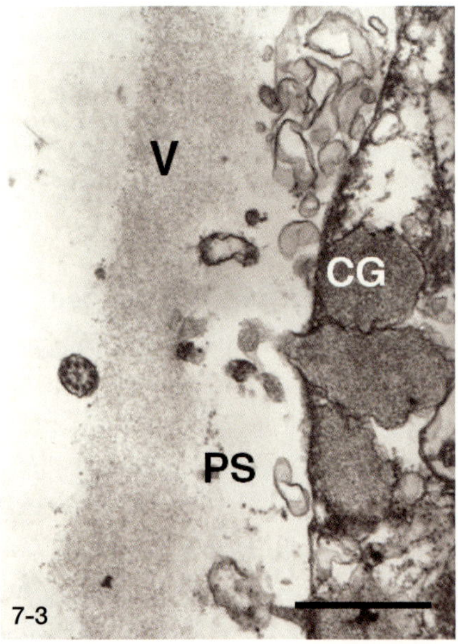

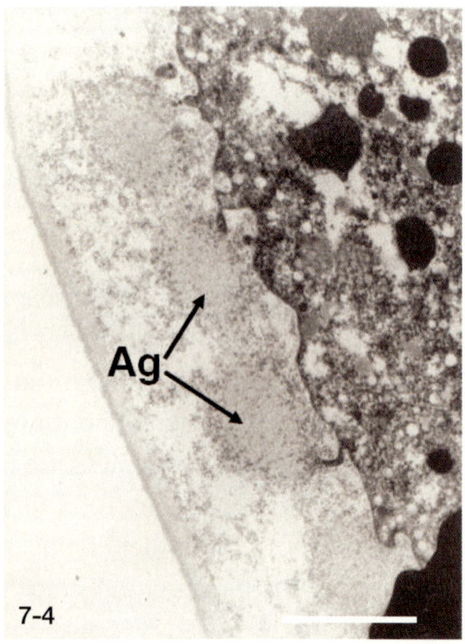

Fig. 7-3 The secretion of the cortical granules (CG) is delayed in *U. unicinctus*. At 4~5 min after the sperm oocyte interaction, the cortical granules align the oolemma beneath and begin to secrete into the elevated perivitelline space (PS). V: vitelline layer, Bar: 2 μm.

Fig. 7-4 After the secretion, the CGs are transformed into a swelled form of an aggregate (Ag) diffusing marginal substance into the perivitelline space. Bar: 2 μm.

solved the wide adhesive area of the oocyte-sperm interaction. The acrosomal process (500 nm in length) projected from the central hole of the acrosome into the oolemma, and the actomere fused with the microvillous tip of the oocyte (Fig. 7-2).

Due to the small and narrowed contact area of the thin microvilli and acrosomal tips, the initial binding is not appear to be stable enough for the subsequent fusion of sperm and oocytes. The stability of sperm-oocyte adhesion is assured by the subsequent formation of spacious adhesion between the compatible curvature of the massive arosome and vitelline layer of the oocyte. As the sperm approached the oocytes, the marginal site of the acrosome simultaneous reached the vitelline layer along with the acrosomal processes. This is due to the concave curvatures of the acrosome, which extended toward the front that are as long as the extensions of the

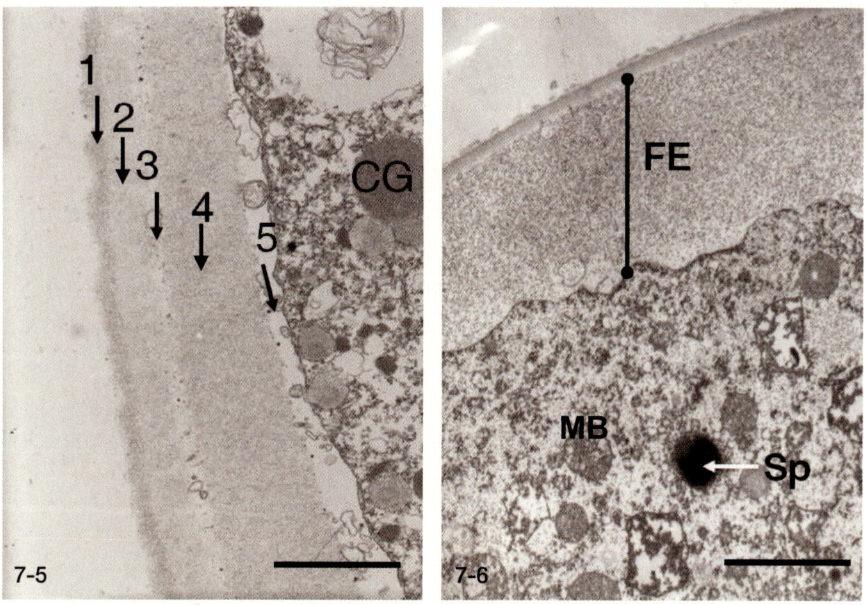

Fig. 7-5 In the continuations of the CG diffusion, the secreted CG materials are fused one another to form larger aggregates (Ag) dividing the fertilization envelope into several layers. The layer differention is numbered. Bar: 2 μm.

Fig. 7-6 At the time of the fertilized spermatozoon (SP) is infiltrated into the oocyte cortex, the fertilization envelope (FE) is differentiated finally into two layers. The inner homogeneous layer is based on the CG secretions and the outer layer is derived from the vitelline layer of the oocyte. MB: multivesicular body. Bar: 2 μm.

marginal sites of the acrosome (Fig. 7-2). The concave frontal acrosome ensured that the adhesion area is at least 1 × 2 μm^2, approximately the size of the sperm head.

2) Formation of the itFE

The itFE is primarily composed of CG secretions, and is defined as differentiated FE containing CG secretions. About 5 minutes after fertilization when the vitelline layer is about 15 μm thick, the CGs are translucent and aligned with the oolemma beneath. The membranes of the CGs fused with the oolemma which eventually burst at the fused sites. The entire content of the CGs are pinched off from the oocyte into the thickened vitelline layer. CG secretions are not humoral overflowing into the vitelline layer, but resembled a suspended solid material that maintained the granular configuration. Secreted CGs in the vitelline layer (Fig. 7-4) are slightly larger (1.5 - 2 μm) in size compared to the original cytoplasmic CGs (1 μm in diameter). Over time, CG secretions in the vitelline layer gradually enlarged or swelled, and diffused from the marginal site. The diffusions intermingled with humoral substances of the iFE and formed various layers. The layers are separately visible over time. The visible layers are numbered inwardly for convenience; the outermost layer is layer 1 and the innermost layer is layer 5. The material source of layers 1 and 2 is the surface coat of the iFE. Substances for the formation of layers 3~5 are derived from CG secretions. These secretions dispersed outward from the itFE, while the surface coat dispersed inward, separating into two layers according to optical densities. The in- and outward diffusion of the surface coat and CG secretions resulted in five layers of the itFE within 8 min following the addition of sperm (Fig. 7-5).

3) Formation of the fFE

The fFE is an elaborate extracellular layer of the oocyte primarily composed of CG secretions and humoral substance previously contained in the iFE. Diffusion and intermixture of the itFE materials continued until layers 2~5 are completely combined into a homogenous monolayer (~14 μm thick). The thick homogeneous monolayer later became the inner layer of the fFE, while layer 1 developed into the thin outmost layer of the fFE. At 15 min after fertilization, the inner fFE appeared to be hardened according to the reflected electron images, in which the electron reflexes are similar to that of semisolid and elastic materials. The fine particles are uniformly distributed all over the inner fFE. Consequently, it can be concluded that CG secretions are necessary for formation of the fFE. In addition, sub-

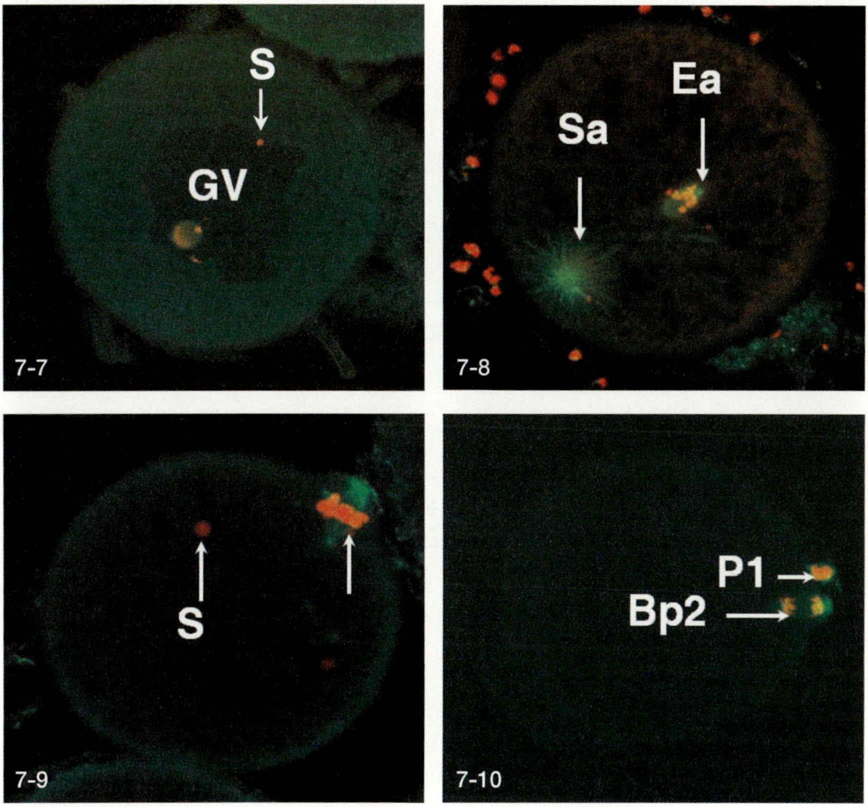

Fig. 7-7 Until 4 min of sperm addition, the CLSMs reveal the intact germinal vesicle (GV), even though the oocyte is fertilized. S: fertilized spermatozoon. × 400.
Fig. 7-8 At completion of the 2nd meiosis, sperm monaster (Sa) begin to activated, while oocyte monaster is stayed inactive. × 400.
Fig. 7-9 The fertilized spermatozoon (S) remains in the ooplasm until completion of the first meiosis. Arrow: the first meiotic apparatus. × 400.
Fig. 7-10 During the successive 1st and 2nd meiosis, three sets of the chromosome are removed remaining one chromosomal set in the ooplasm. At the completion of the 2nd meiosis, the monaster of the fertilized spermatozoon begin to activate. P_1: the first polar body, BP_2: bipolar meiotic apparatus. × 400.

stances in the perivitelline space played important role as a solvent for the CG secretions. The substance of the perivitelline space should be analyzed to understand the mechanism for preventing polyspermy and fFE formation.

During fFE formation at 7-8 min, germinal vesicle break down (GVBD) is completed, and the oocyte nucleus entered the first meiotic phase following fFE formation. Soon after the extrusion of the first polar body, the oocyte entered the second meiosis. The spermatozoon resided in the ooplasm until the second polar body extruded into the fFE (Fig. 7-7, 7-9). The monaster of the spermatozoon is activated when the fertilized oocyte entered into the second meiotic phase. Later, the activated sperm monaster initiated the fusion of both pronuclei (Fig. 7-8).

7-3 Cortical reactions and polyspermy block in *U. unicinctus* oocyte

It is known in many marine invertebrates, CGs are secreted soon after sperm-oocyte interaction, and the secretions cause the elevation of the vitelline layer to prevent polyspermy. This occurs when CG secretions overflow onto the oolemma. The cortical reaction, formation of FE, and polyspermy prevention in *U. unicinctus* are differed from other organisms in many respects (Shin *et al.*, 2005). First, the timing of CG secretion is differed from that of other invertebrates. CGs are secreted into the raised perivitelline space, do not cause the elevation of the perivitelline space because secretion is delayed until the perivitelline space is completely raised. Second, the CG secretion does not overflow onto the oolemma, but rather maintains granular configuration slightly larger than the CGs found in the ooplasm. Third, the granular secretions gradually disperse in the perivitelline space to form the homogeneous monolayer of the fFE. Fourth, the substance previously contained in the perivitelline space seems to play role as a solvent for the CG secretions. Fifth, the massive acrosomal materials derived from the fertilized spermatozoon are dispersed in the surface coat of the fFE, seem dedicated to formation of the FE. In addition, because of the delayed secretion of the CGs and the short half-life of the sperm, CG secretions seem not primarily dedicated to preventing polyspermy. The vitelline layer and perivitlline space are the effective barriers against polyspermy until CGs are secreted. Hence, FE formation and polyspermy blocking in *U. unicinctus* are not completed simultaneously with the cortical reaction. Formation of the FE is subdivided into three phases according to secretion, dispersion, and refinement of the CG materials as described above.

In different animal species, it is known that polyspermy is prevented by restricting the number of sperm that approach the oocyte, either by narrowing the micropyle or developing an extracellular coat such as an FE or zona pellucida (Wolf *et al.*, 1981). Another mechanism for blocking polyspermy is restricting the number of sperm pronuclei that combine directly with the oocyte pronuclei when polyspermy occurs (Jaffe & Gould-Somero, 1985). As is described above, *U. unicinctus* oocytes can be categorized into the group that develops an extracellular coat for preventing polyspermy similar to other marine invertebrates. The extracellular coat, commonly known as the FE in marine invertebrate, is composed of highly differentiated extracellular structures including the fertilization cone, vitelline layer (Anderson, 1968), glycocalyx (Bennett, 1963), perivitelline space, and CG secretions.

There are many animals in which CGs are observed in the ooplasm, and CG secretions cause the vitelline layer to expand, leading to the formation of the FE. However, there are various opinions on the involvement of CGs in FE formation. The contents of the CGs are secreted into the vitelline layer to establish the FE in echinoderms (sea urchin: Vacquire, 1975; Eddy & Shapiro, 1976; Berg & Wessel, 1997; Fisher & Rebhun, 1983), star fish (Sardet & Chang, 1987), mollusks (Speksnijder *et al.*, 1986), ascidians (*Phallusia mammillata*; Sardet *et al.*, 1994), annelids (*Tubifex hattai*; Shimizu, 1995), amphibians (*Xenopus*; Larabell, 1993; Chang *et al.*, 1999; Nuccitelli & Ferguson, 1994; Gerhart *et al.*, 1989), and mammals (mouse; Wasserman, 1977, 2001). In these organisms, polyspermy is prevented by two sequential blocking mechanisms. Early blocking is followed by the cortical reaction, which is the causal event for the late blocking of polyspermy in participation of CG secretions (Rothschild & Swann, 1952). On the contrary, CGs are released prior to insemination in *Mytilus edulis*;

hence, there is no cortical reaction for initiating the late polyspermy blocking and the FE is not formed (Pasteels *et al.*, 1958; Pasteel, 1965; Humphreys, 1962). Similarly, FE raise is triggered prior to insemination by exposure to seawater in *Sabellaria* (Pasteels, 1965). In *Spisula*, CGs are not released in response to insemination, but polyspermy is rapidly prevented (Rebhun, 1962). In mammals, CG secretion occurs without significant raise of the zona pellucida (Szollos *et al.*, 1967). Based on these findings, it is apparent that the prevention of polyspermy based on CG secretion is accomplished via mechanisms that differ from species to species.

Difference in the appearances and behavior of CGs and CG secretion is also seen among *Urechis*. In the literature, different detailed procedures of FE formation and polyspermy block have been described. Some authors noted that CGs are not found in *U. caupo* (Gould-Somero & Hollander, 1975b), while others reported that CGs are present but not participate in FE formation (Gould-Somero & Hollander, 1975a). The former authors proposed that the FE fails to block polyspermy, while the latter authors found that the FE is effective for prevent polyspermy even though the CGs are not associated with FE formation. In this case, the number of CGs is unchanged and the CGs maintain their original configuration until the fertilized egg develops into an embryo (Gould-Somero & Hollander, 1975a). It is thought that in *U. caupo* the mechanism for polyspermy blocking appears to be fundamentally different from that of other animals (Gould-Somero & Hollander, 1975a). Consumption of CG secretions during formation of the FE and prevention of polyspermy differs from species to species (Barros & Yamaguchi, 1971; Navara & Schatten, 1994). As above, various groups have observed different processes for cortical reaction, FE formation, and polyspermy prevention.

Fertilized *U. unicinctus* oocytes develop FEs derived from massive CG secretions (Kwon *et al.,* 2000, 2003: Shin *et al.,* 2005). Under the conditions of the fertilization experiments, activation of the CG secretions and subsequent procedures require some time to convert the thickened itFE containing CG secretions into the fFE. The FE formation in *U. unicinctus* oocyte is based on CG secretions as the results of the cortical reactions given an impulse of the sperm. Furthermore, the materials derived from the acrosome of the fertilized spermatozoon are intermixed with the surface coat, which is the most powerful barrier against polyspermy (*See* Acrosome).

CG secretions are completely dispersed and rearranged into a homogeneous monolayer of the inner fFE. During fFE formation, the humoral substances that previously filled the perivitelline space serve as a solvent for facilitating the diffusion and rearrangement of the CG secretions. This is because the CG secretions are suspended solid matter rather than humoral substances overflowing onto the oocyte surface. Hence, the capacity and character of the perivitelline space contents are important factors for fFE formation. Diffusion and differentiation of CG secretions and rearrangement according to gradations of the materials are based on the characteristics of the iFE substances. For the reason, each layer is combined and ultimately forms two layers of the fFE: the thin outer surface coat containing a small amount of acrosomal materials, and a thick homogeneous inner monolayer. The inner layer is later hardened to construct a permanent barrier against polyspermy, as reported in *Echinoderm* (Foerder & Shapiro, 1977). Although FE is formed in participation of CG secretions, the FE seems little dedicated to prevent polyspermy, because the FE formation is distinctly delayed. At completion of the FE, no sperm are competent to interact with ooyte. As is known in other animals, the FE in *U. unicinctus* has to play different roles: one is the formation of a permanent barrier to prevent polyspermy, and the other is to provide a space in which future embryos continue to develop (Shin *et al*., 2005).

In *U. unicinctus*, the cortical reaction is delayed by 5 min from the time of sperm addition, and seemed not be directly involved in preventing polyspermy. The half-life of sperm is 1~2 min when exposed in NSW. For the while, the sperm are vigorously active and devoted their energy to participate in fertilization. The sperm receptors of the oocyte are intact until the first sperm-oocyte contact, which induce microvilli retraction into the raising vitelline space. Cortical reactions did not occur during the critical time periods. Instead, the iFE or the raising of the vitelline layer blocked polyspermy and are accompanied by retraction of microvilli, which spread out into the oolemma to compensate for the increased oocyte volume. It can be inferred that the raised iFE forms a unique barrier to prevent multiple rounds of fertilization before occurrence of the cortical reaction. The raise of iFE forces out surplus sperm from the oolemma until the iFE is 15 μm thick about 5 min after sperm addition. Along with structural change, changes of sperm receptor quality in *U. caupo* (Gould & Stephano, 1993) and echinoderms (Sato & Osanai, 1986; Barros & Yangimachi, 1971; Carroll & Epel, 1975) are achieved through the mechanism for polyspermy blocking.

To demonstrate the effectiveness of the iFE, the vitelline layer of the oocyte is removed using 5% Triton X-100 prior to fertilization. After being carefully wished several times, the nude oocytes are artificially fertilized and treated under the conventional method and conditions. The sperm used for fertilization are visualized by Propidium iodide fluorescence staining. Under CLSM, the individual fluorescent sperm could be counted. In contrast to oocytes with an intact vitelline layer, polyspermy is greatly increased among nude oocytes. A maximum of 50 sperm are counted in the cytoplasm of the nude oocyte. The differences between the intact and nude oocytes provide evidence for the role of the vitelline layer during fertilization. On the other hand, when the oocytes are immersed in sea water of lower salinity, the vitelline layer is elevated prior to fertilization. These oocytes are not easily fertilized, based on the report that fertilization should occur within 1 min of adding sperm in *U. caupo* (Paul, 1975).

In addition, Propidium iodide fluorescence staining revealed the frequency of polyspermy in the cytoplasm of fertilized oocytes. The fate of extra spermatozoon in the normal fertilized oocytes is not determined. It is has been tentatively suggested that all of polysperm oocytes do not develop into abnormal embryos. It is possible that spontaneous polyspermy can occur under natural conditions in *U. unicinctus*, and restriction mechanism on the level of nuclear fusion parallel the restriction of sperm numbers at the intracellular level (Jaffe & Gould-Somero, 1985).

Summary

It is known in many animal species, FE is developed on substantial bases of CG secretions, plays role in permanent block to polyspermy and preparing extracellular space for the fertilized oocyte proceeding prospective cleavage and further embryonic developments. The situations in concern with CGs are similar to *U. unicinctus* oocyte, although there are various different opinions on the presence and function of the CGs in *Urechis* oocyte.

Two characteristics of the CG secretion are observed: The CGs of *U. unicinctus* are not secreted soon after the sperm-oocyte interaction, but delayed until vitelline layer is completely elevated. The raise of vitelline layer takes time, the CGs are discharged into the thickened perivitelline space. The other feature is contour of the secretions. The CG secretion in *U. unicinctus* is not humoral, but semisolid suspension. Globular contour lines of the CGs are maintained in the perivitelline space, even though the CGs have been secreted. Thereafter, the CGs are dispersed until homogeneous monolayer is formed in the perivitelline space. For a while of the oocytes are exposed to numerous sperm, the vilelline layer and perivitelline space fill role in blocking polyspermy. In *U. unicinctus,* the blocking role to polyspermy of the completed FE is exclusive for the reason of the delayed CG secretion. The thick FE contained CG secretions is effective providing extracellular space for the prospective embryonal development.

In addition, the vitelline layer of oocye is previously removed and fertilized to study role of outer acrosomal material of spermatozoon. The outer acrosomal materials are not consumed for a considerable period of time adhering on the oolemma. The most of the "nude oocytes" is multiple fertilized, polysperm in the nude oocyte are revealed on fluorescence microscope.

REFERENCES

Anderson E: Oocyte differentiation in the sea urchin, *Arbacia puncrulata*, with particular reference to the origin of cortical granules and their participation in the cortical reaction. J Cell Biol, 37: 541-538, 1968.

Barros C, R Yanagimachi: Induction of zona reaction in golden hamster eggs by cortical granule material. Nature, 233: 268-269, 1971.

Bennett HS: Morphological aspects of extracellular polysaccharides. J Histochem Cytochem, 11: 1-6, 1963.

Berg LK, GM Wessel: Cortical granules of the sea urchin translocate early in oocyte maturation, Development, 124: 1845-1850, 1997.

Carroll EJ, D Epel: Isolation and biological activity of the protease released by sea urchin eggs following fertilization. Dev Biol, 44: 22-32, 1975.

Chang P, D Perez-Mongiovi, E Houliston: Organization of *Xenopus* oocyte and egg cortices. Microsc Res Tech, 44: 415-429, 1999.

Conner S, D Leaf, G Wessel: Members of the SNARE hypothesis are associated with cortical granule exocytosis in the sea urchin egg. Mol Reprod Dev, 48: 106-118, 1997.

Eddy EM, BM Shapiro: Changes in the topography of the sea urchin egg after fertilization. J Cell Biol, 71: 35-48, 1976.

Fisher GW, LI Rebhun: Sea urchin egg cortical granule. Exocytosis is followed by a burst of membrane retrieval via uptake into coated vesicles. Dev Biol, 99: 456-472, 1983.

Foerder CA, BM Shapiro: Release of ovoperoxidase from sea urchin eggs is hardend the fertilization membrane with tyrosine crosslinks. Proc Natl Acad Sci *USA,* 74: 4214-4218, 1977.

Gerhart J, M Danilchik, T Doniach, S Robert, B Bowning, R Stewart: Cortical rotation of the Xenopus egg: Consequence for the anterioposterior pattern of embryonic dorsal development. Development: 37-51, 1989(suppl.).

Gould -Somero M, JL Stephano: Nuclear and cytoplasmic pH increase at fertilization in *Urechis caupo.* Dev Biol, 159: 608-617, 1993.

Gould-Somero M, L Hollander: Fine structural investigation of the insemination response in *Urechis caupo.* Dev Biol, 46: 358-369, 1975a.

Gould-Somero M, L Hollander: Oocyte differentiation in *Urechis caupo* (Echiura): A Fine structural study. J Morphol, 147: 475-505, 1975b.

Humphreys WJ: Electron microscope studies on eggs of *Mytilus edulis.* J Ultrastruct Res, 7: 467-487, 1962.

Humphreys WJ: The fine structure of cortical granules in eggs and gastrulae of *Mytilus edulis.* J Ultrastruct Res, 17: 314-326, 1967.

Inoue S, LG Tilney: Acrosomal reaction of *Thyone* sperm: I. Change in the sperm head visualized by high resolution video microscope. J Cell Biol, 812-819, 1982.

Jaffe LA, M Gould-Somero: Polyspermy-penetrating g mechanism. In: Biology of Fertilization. Vol. 3, Academic Press, New York pp 223-250, 1985.

Kwon HJ, KS Shin, WJ Kim: Fine structural investigation of fertilization envelope and acrosomal reaction in *Urechis unicinctus*. Korean J Electron Microscopy, 30(1): 61-72, 2000.

Kwon HJ, JW Jeong, WJ Kim KS Shin: Immunofluorescence and electron microscopic study on the artificial insemination and rotation-shift behaviors of the bipolar spindle fibers in *Urechis unicinctus* egg. Korean J Electron Microscopy, 33(2): 105-116, 2003.

Larabell CA: A new technique for isolation and visualization of the *Xenopus* egg cortex reveals a complex cytoskeleton. J Struct Biol, 110: 154-159, 1993.

Navara CS, First NL, Schatten F: Microtubule organization in the cow during fertilization, polyspermy, parthenogenesis, and nuclear transfer: The role of the sperm aster. Dev Biol, 29-40, 1994.

Matese JC, S Black, DR McClay: Regulated exocytosis and sequential construction of the extracellular matrix surrouding the sea urchin zygote. Dev Biol, 186: 16-26, 1997.

Nuccitelli R, JE Ferguson: Fertilization and ion channels. Science, 263: 998, 1994.

Pateels JJ, P Castiaux, G Vandermursscche: Cytochemical localizations and ultrastructure in the fertilized unsegmented egg of *Paracentrotus lividus.* J Biophys Biochem Cytol, 4: 575-577, 1958.

Pasteels JJ: Etude au micoscope electronique de la reaction corticale. II. La reaction corticale de l'oeuf vierge de *sabellaria alveolata.* J Embryol Exp Morphol, 13: 327-329, 1965.

Paul M: The polyspermy block in eggs of *Urechis caupo*: evidence for a rapid block. Exp Cell Res, 90: 137-142, 1975.

Rebhun LI: Electron microscope studies on the vitelline membrane of the surf clam, *Saisula solidissima.* J Ultrastruct Res, 6: 107-122, 1962.

Rothschild L, MM Swann: The fertilization reaction in the sea urchin. The block to polyspermy. J Exp Biol, 29: 469-483, 1952.

Sardet C, M Chang: The egg cortex: from maturation through fertilization. Cell Differ, 21: 1-9,

1987.

Sardet C, A McDougall, E Houliston: Cytoplasmic domains in egg. Trends Cell Biol, 4: 166-172, 1994.

Sato M, K Osanai: Morphological identification of sperm receptors above egg microvilli in the polychaete, *Neanthes japonica.* Dev Biol, 113: 263-270, 1986.

Shimizu Y: Role of the cytoskeleton in the generation of spatial patterns in *Tubifex* eggs. Curr Top Dev Biol, 31: 197-235, 1995.

Shin KS, DH Lee, TY Ko: Cleavage pattern of *Urechis unicinctus* egg in vitro fertilization. Korean J Electron Microscopy, 34(1): 71-81, 2004.

Shin KS, HJ Kwon, WJ Kim: Attribution to cortical granules to formation of fertilization envelope and polyspermy block in *Urechis unicinctus*. Integrative Biosciences, 9(2): 57-64, 2005.

Speksnijder JE, WA de Jong K, Linnemans, MR Dohmen: The ultrastructural organization of the isolated cortex of a molluscan egg. Prog Clin Biol Res, 217B: 353-356, 1986.

Stricker SA: Comparative biology of calcium signaling during fertilization and egg activation in animals. Dev Biol, 211: 157-176, 1999.

Szollos D, P Calarco, RP Donahue: Absence of centrioles in the first and second meiotic spindles of mouse oocyte. J Cell Sci, 11: 521-541, 1967.

Tilney LG, S Inoue: Acrosomal reaction of the Thyone sperm. III. The relationship between actin assembly and water influx during the extension of the acrosomal process. J Cell Biol, 100: 1273-1283, 1985.

Vacquier VD: The isolation of intact cortical granules from sea urchin eggs: calcium ions trigger granule discharge. Dev Biol, 43: 62-74, 1975.

Vacquier VD: Dynamic changes of the egg cortex. Dev Biol, 84: 1-26, 1981.

Wasserman PM, TE Ukena, WJ Josefowicz, GE Letourneau, MJ Karnovsky: Cytocalasin B-induced pseudocleavage of mouse oocytes *in vitro.* J Cell Sci, 26: 323-337, 1977.

Wasserman PM, L Jovine, ES Lischer: A profile of fertilization in mammals. Nat Cell Biol, 3: E59-64, 2001.

Wolf DE, W Kinsey, WJ Lennarz, M Eddin: Changes in the organization of the sea urchin egg plasma membrane upon fertilization: Indications from the lateral diffusion rates of lipid soluble fluorescent dyes. Dev Biol, 81: 133-138, 1981.

Chapter 08

Artificial fertilization and parthenogenetic study on the origin of the meiotic apparatus in *Urechis* oocytes

Chapter **08**

Artificial fertilization and parthenogenetic study on the origin of the meiotic apparatus in *Urechis* oocytes

8-1 Introduction

In many animal species including mammals (Lopata, *et al.*, 1988), the most obvious event of oocyte maturation is germinal vesicle break down (GVBD), which indicate the end of prophase I of meiosis. After GVBD, maturation of the oocyte is stagnated or arrested at a stage between prophase (G2 phase) of the first meiosis and metaphase of the second meiosis (Schultz, 1967; Colas & Dube, 1998). Fertilization triggers further developments of the oocyte, and the oocyte is extricated from the meiotic arrest. Some animal species is differed with fertilization time from the most of animal species; some Coelenterates and Echinoid (sea urchin) are fertilized at post-meiotic pronuclear stage, and some annulate worm, flatworm and polychaete are fertilized at pre-meiotic stage with intact germinal vesicle (Brachet, 1991).

A number of literatures report on the meiotic arrest and re-initiation of oocyte developments following on fertilization. The fertilization initiates activation or inactivation of cytoplasmic factors including maturation promoting factors (MPF; Masui, 1991; Segata, 1996; Whitaker, 1996; Taieb *et al.,* 1997), maturation inhibiting factor (MIF) and cytostatic factor (CSF, Colas & Dube, 1998; Kishimoto, 1999). By a chance of fertilization, the oocyte continues further developments overcoming the meiotic arrest. Re-initiation of the oocyte developments are closely related to bring the stagnated meiotic apparatus into action.

The meiotic apparatus (Map) is derived from pre-mitotic aster-like structure (PMA) appeared in cytoplasm of pre-meiotic unfertilized oocyte (Miyazaki, *et al.*, 2000; Uetake, *et al.*, 2002). The PMA is developed into Map during oocyte maturation, and further meiotic progression is triggered by fertilization, as is described above.

Urechis oocyte is fertilized prior to GVBD; young primary oocyte with intact germinal vesicle is fertilized encountering meiosis (Gould-Someros, 1975; Gould-Someros & Hollander, 1975, Kwon *et al.*, 2003; Shin *et al.*, 2004). Soon after occurrence of fertilization, germinal vesicle is break-downed, and the oocyte enters meiotic phase. GVBD is not consequential event of oocyte maturation, but an indication ascertaining the occurrence of fertilization. After fertilization, events of fertilization and maturation including GVBD and meiotic cycles are continued without a stagnated phase until the female and male pronuclei are fused.

In order to observe origin of Map in *Urechis* oocyte, unfertilized and parthenogenetic oocyte were stained with anti-tubulin-FITC according to Bestor & Schatten (1981) and Propidium iodide followed by confocal laser scanning microscopy (CLSM) and electron microscopy (EM). In proceeding of GVBD, immunofluorescence active particle (IAP) was observed in nuclear region of the parthenogenetic oocyte. During meiosis of the parthenogenetic oocyte, the Map of *Urechis* oocyte was derived from the IAP. In the sense, the IAP is analogous to the PMA of other animal species. Site and time of IAP appearance were differed from the site and time of PMA appearance. As the origin of Map in *Urechis* oocyte, the IAP was compared with PMA of other animal species.

In the course of the observation, fertilized oocyte revealed the whole process of meiosis and behaviors of the meiotic apparatuses. During meiosis of fertilized oocyte in *U. unicinctus*, migration and "rotation-shift" of the Map as well as successive elimination of active centriole were observed. The meiotic phenomena are compared with the meiotic events progressed in parthenogenetic oocyte.

8-2 Artificial fertilization

In brackwash water, cytoplasmic indentation was gradually disappeared, the mature oocyte restored round shape exposing microvilli, sperm receptors, on the entire surface of the oocyte. Prominent germinal vesicles (GV, 50 μm in diameter) and the nucleolus (3 μm in diameter) were clearly observed under conventional LM (Fig. 8-1a, b). Appropriate amount of sperm suspension was added in the oocyte suspension, or oocyte suspension was added in the sperm suspension to induce artificial fertilization. The latter method was in practice convenient. Within 1 min of oocyte addition, the vitelline layer was raised to form a thickened perivitelline space through which microvilli (sperm receptors) protruded beyond the vitelline layer. The perivitelline space and vitelline layer was a reliable barrier for preventing polyspermy. Propidium iodide staining followed by CLSM revealed numerous surplus sperm withdrawing from the raised vitelline layer (Fig. 8-3, 8-19).

After 2 min of oocyte addition, the membranes of the spermatozoon and oocyte fused; the sperm-oocyte interaction was observed on sections of the median plane of the spermatozoon. The entire frontal surface of discoid ac-

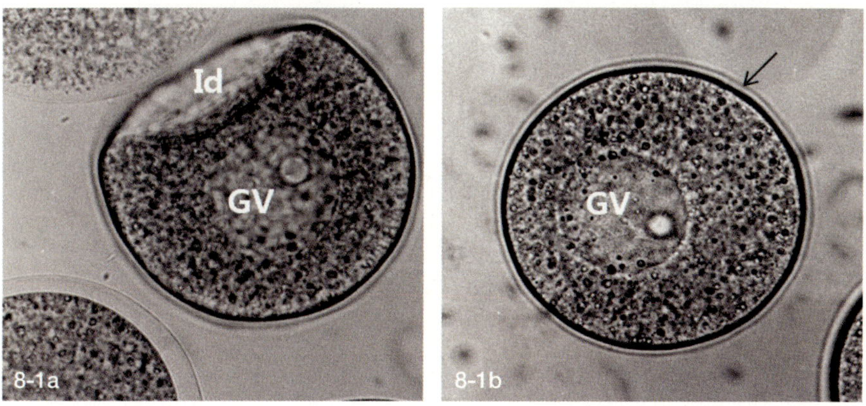

Fig. 8-1 (a) Cytoplasmic indentation (Id) and nuclear invagination are characteristics in the mature unfertilized *Urechis* oocyte. GV: germinal vesicle containing nucleolus. × 700. (b) Natural sea water in the habitat is brackish water of lower salinity. After spawning in the brackish water, the oocyte restores round shape recovering the indentation and invagination. Arrow: vitelline layer, GV: germinal vesicle. × 700.

rosome was contacted on the vitelline surface assuring the wide range of sperm-oocyte contact area (Fig. 8-2a). Spermatozoon was not always arranged perpendicular to oolemma at the sperm-oocyte interaction. It was observed many sperm approach oocyte parallel to the oocyte surface or with an acute angle (Fig. 8-2b). The short acrosomal process (500 nm in length) was fused to the microvillous tip while other microvilli were retracted into the perivitelline space (Fig. 2a). In addition, germinal vesicle was break-downed (GVBD) at around 2 min of the fertilization (Fig. 8-3, 8-4).

About 5 min after fertilization, the thickness of the perivitelline layer was maximized to 15 μm (Fig. 8-4). CGs were eventually secreted into the perivitelline space. Initially, the secretion was not humoral overflowing the oolemma, but suspended solid (1.5~2 μm) maintaining granular configuration which slightly larger than the original CGs (1 μm) found in the oo-

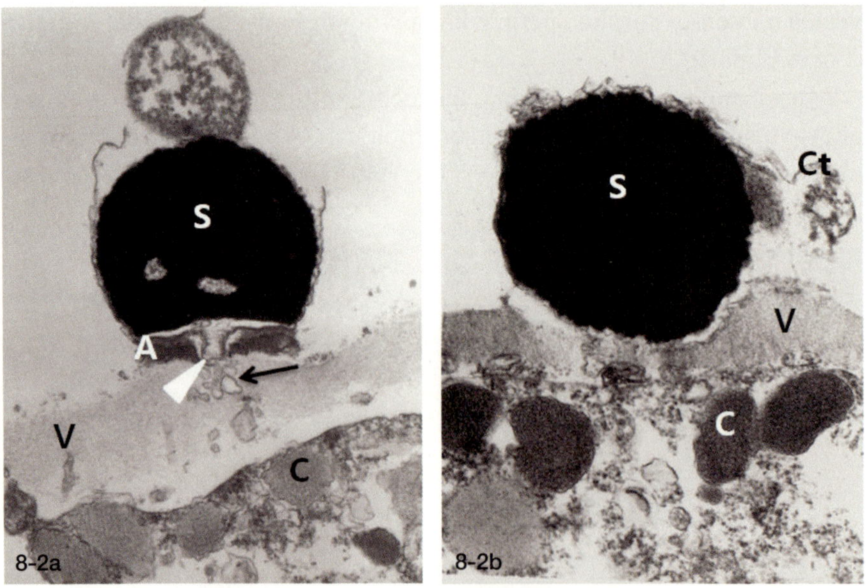

Fig. 8-2 (a) Acrosomal process of spermatozoon (white arrow head) fuses with oocyte microvillus (black arrow). The outer acrosomal material is intact, and not consumed adhering on surface of vitelline layer. A: outer acrosome, C: cortical granule, S: spermatozoon, V: vitelline layer, × 8,000. (b) Spermatozoon is not always approached perpendicular to oocyte surface as seen in the figure. It is likely that the oocyte is active, not stayed still in interaction with the spermtozoon. C: cortical granule, Ct: centriole of the fertilizing spermatozoon, S: spermatozoon, V: vitelline layer. × 12,000.

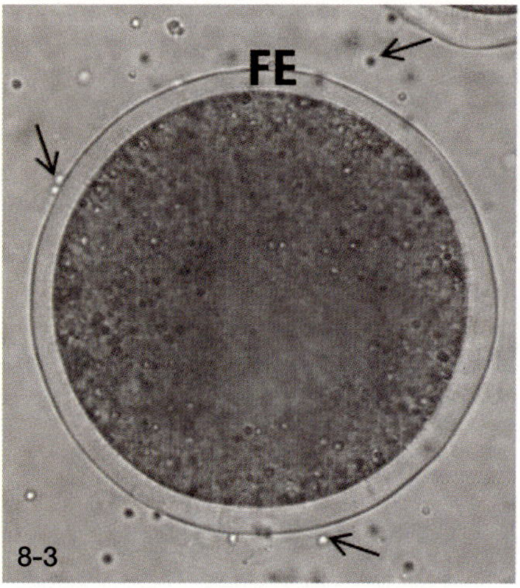

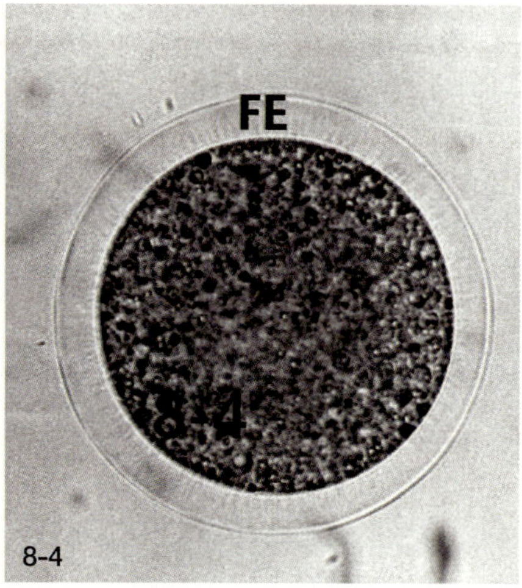

Fig. 8-3 During formation of fertilization envelope (FE) at 2 min of the fertilization, germinal vesicle is break-downed. The cloudy portion in middle of oocyte is nuclear region where the germinal vesicle is previously break-downed. The FE forces surplus sperm (arrows) distant out of the oocyte. × 800.

Fig. 8-4 At 5 min of the fertilization, the FE is maximized in thickness including retracted microvilli observed as reflections of radial arrays in the FE. The nuclear region is faded out intermixing nucleoplasm and ooplasm. × 700.

cyte cortex. The suspended solid secretions of CGs were gradually disappeared in the perivitelline layer to form homogeneous layer of the final fertilization envelope (fFE) (*See* Cortical granules). At 40 min and 90 min of the sperm addition, the first and second polar bodies were squeezed out of the oocyte respectively (Fig. 8-5, 8-6). At the second meiosis, division of the first polar body was observed (Fig. 8-17, 8-18).

8-3 Immunofluorescence staining and meiosis

8-3-1. First meiosis

Prior to GVBD, the diploid chromosome set was grouped and arranged near the nuclear envelope (Fig. 8-11). Before occurrence of GVBD, the fer

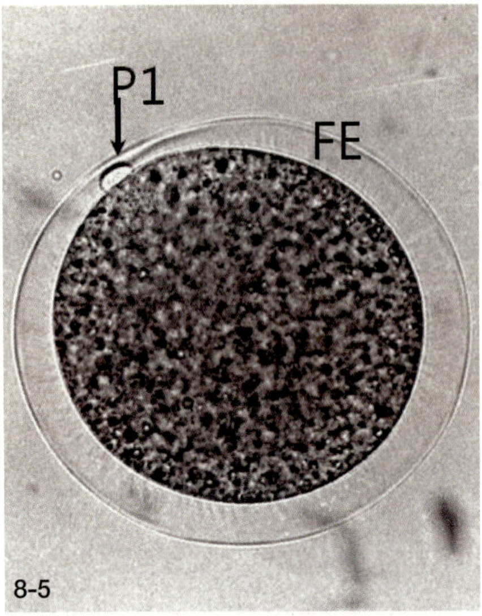

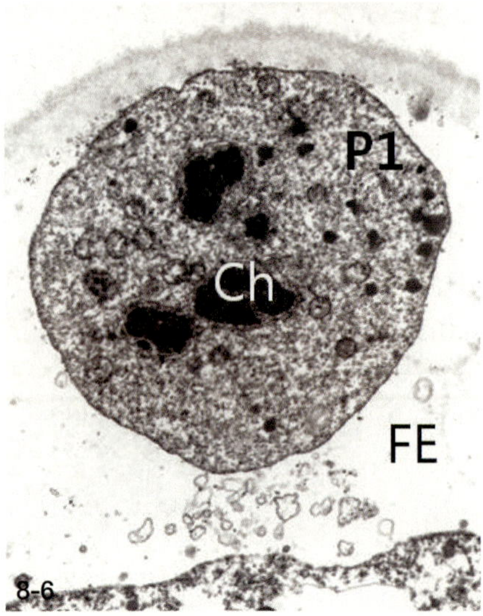

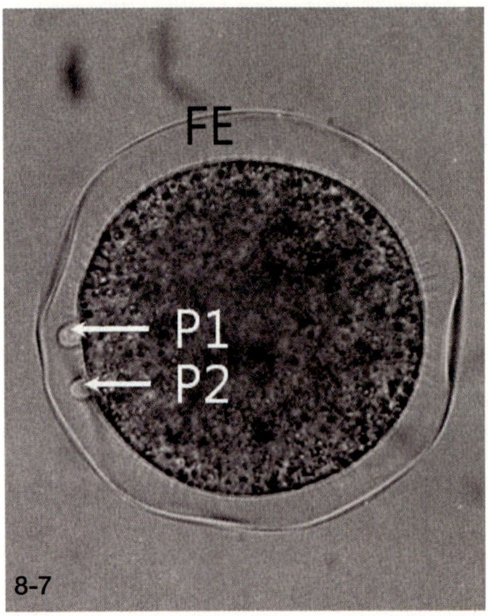

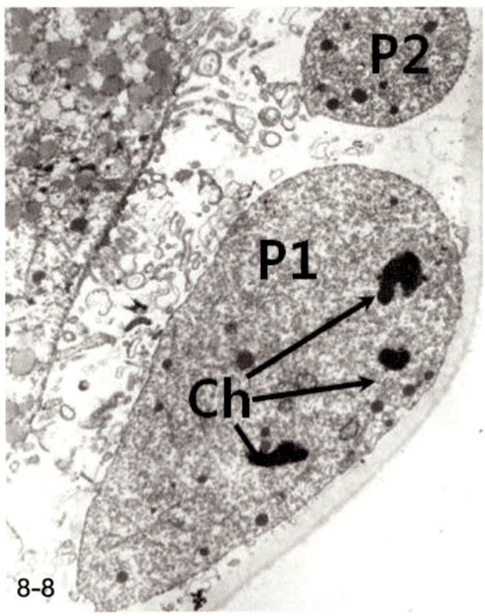

Fig. 8-5 At near 40 min of the fertilization, the first polar body (P1) is extruded into the fertilization envelope (FE). × 700.

Fig. 8-6 The first polar body (P1) in the FE contains one chromosome set (Ch) and an amount of cytoplasm. × 20,000.

Fig. 8-7 At 90 min of the fertilization, the second polar body (P2) is squeezed out into the FE at near of the P1 site. × 700.

Fig. 8-8 The P1 is larger than P2, and the site of P2 approaches the site of P1. The P1 is subsequently activated producing the third polar body. Ch: chromosome set of P1. × 20,000.

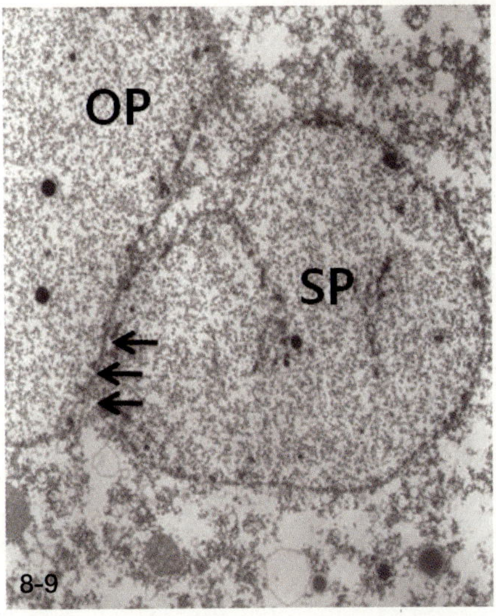

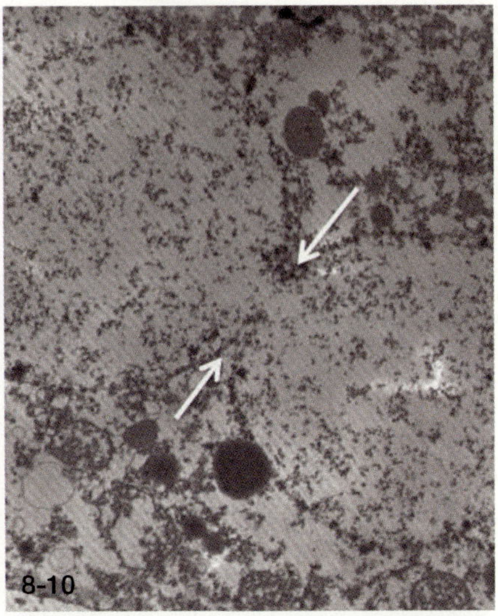

Fig. 8-9 At an initial phase of pronuclei fusion, male pronucleus (SP) is distinguished from female pronucleus (OP) by infoldings of the nuclear envelope. Arrows: contact site of sperm and oocyte pronuclei. × 24,000.

Fig. 8-10 Male and female pronuclei are fused at the constricted site (arrows) between both pronuclei. × 24,000.

tilized spermatozoon was found in common at eccentric ooplasm between the oolemma and germinal vesicle, sometimes the spermatozoon was to observed at near the nuclear envelope of the germinal vesicle. In parthenogenetic activation, the precursor of the Map appeared as immunofluorescence active particle (IAP) at 2~3 min after sperm addition (Fig. 8-21), developed into granule like structure (GLS) at 3~4 min (Fig. 8-24), and bipolarity of the first Map was appeared at 15~20 min of the fertilization. In appearance of the biporarity, the first Map began to migrate, the direction of the migration was estimated to opposite oolemma site of the infiltrated spermatozoon (Fig. 8-14). The opposite oolemma site corresponds prospective site of the polar body extrusion.

After 40 min of the fertilization, anti-tubulin-FITC immunostaining revealed that first Map approaches the destined site of the oolemma, arranged perpendicular to the tangential line of the oolemma (Fig. 8-12, 8-13, 8-14), while the infiltrated spermatozoon was still remained inactive in the eccentric ooplasm opposite to the first Map of the fertilized oocyte. The inner and outer sets of chromosomes grouped separately to the median plane of the Map (Fig. 8-13). The outer pole was close associated with the corresponding oolemma, more active in incorporation of the immunofluorescence dyes, observed in the convexly projected portion of the oo-

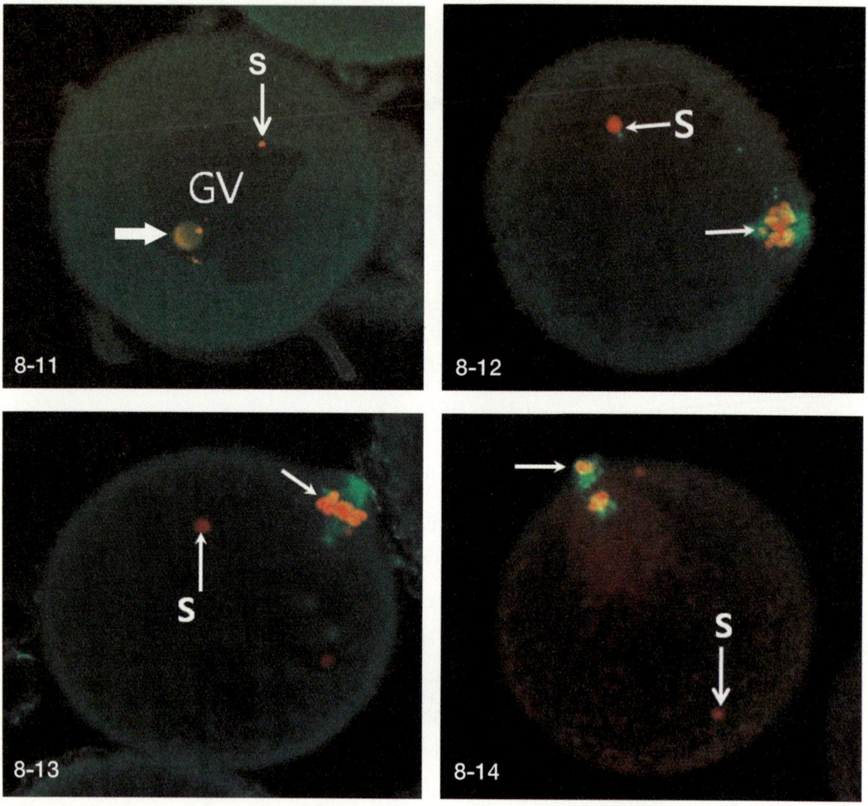

Fig. 8-11 Before GVBD, diploid chromosome set of oocyte (block arrow) is grouped and arranged near the nuclear envelope of the germinal vesicle, while the fertilized spermatozoon (S) is placed at opposite site of the chromosome set. GV: germinal vesicle. × 800.
Fig. 8-12 At around 40 min of the fertilization, the first meiotic apparatus is vertically arranged to the tangential line of the destined site of oolemma. S: fertilized spermatozoon, Arrow: vertical arrangement of the first meiotic apparatus with diploid chromosomal set. × 800.
Fig. 8-13 At the site of oolemma, polarity of the outer centriole is highly active, while the inner centriole is weakened. S: fertilized spermatozoon, Arrow: diploid chromosomal set within the first meiotic apparatus. × 800.
Fig. 8-14 At 40 min of the fertilization, the first polar body (arrow) is extruded into the fertilization envelope. Fertilized spermatozoon (S) is located opposite to the site of the first meiotic apparatus. × 800.

lemma (Fig. 8-13). The first polar body was extruded in subsequent stage from the projected site. At 40 min, the outer chromosomal set extruded from the oocyte to form the first PB in the fFE (Fig. 8-5, 8-6, 8-16).

8-3-2. Second meiosis

During formation and extrusion of the first PB, the inner and outer chromosomal sets were located at near of the destined PB site. After extrusion of the outer chromosomal set from the oocyte, the inner chromosomal set was not stayed at near of the oolemma, but observed initially in the middle portion of the ooplasm. There appeared to be a backward migration of the inner chromosomal set to the inside of the ooplasm.

The remaining chromosomal set in the oocyte entered second meiotic phase. Bipolarity of the second Map was created, arranged parallel to the tangential line of the corresponding oolemma (Fig. 8-15). As it approached the destined oolemma site, the second Map was vertically arranged to the tangential line of the destined oolemma. During migration, the second Map then rotates so that the parallel configuration becomes vertical to the oolemma (Fig. 8-16). The migration pattern of the second Map has been observed in Echinoderm oocytes and termed "rotation-shift" (Sluder *et al,* 1989; Hamaguchi, 2001; Uetake, *et al.*, 2002). The arrangement at appearance of the first Map was not exactly observed in the present study of *U. unicinctus*. However, it was likely that both Maps were shifted and rotated from a parallel configuration to vertical arrangement to the corresponding oolemma. At 90 min following fertilization, the second meiosis cycle was completed with extrusion of the second PB (Fig. 8-16). The extrusion site of the second PB approaches the site of the first PB (Fig. 8-7).

The first PB was spherical (3 μm) immediately after extrusion (Fig. 8-6). As the oocyte entered into the second meiosis cycle, the shape changed into an ellipsoidal figure (5 × 10 μm) in which the elliptic site closely approached the oolemma (Fig. 8-8). In proceeding of the second meiosis cycle, the first PB-spindle fibers were concentrated (Fig. 8-18). Anti-tubulin-FITC and Propidium iodide staining revealed that the first PB-spindle fibers were closely associated with the chromatin components (Fig. 8-17). Eventually, the cleavage furrow of the first PB was observed on the confocal microscope (Fig. 8-17).

During fertilization, it was likely that the spermatozoon discarded mitochondria at close vicinity of the vitelline layer. Spermatozoon nucleus and centrioles were infiltrated into the ooplasm. As described above, the fertilized spermatozoon remained on the site opposite to the Maps during the first meiosis. When the second Map began to migrate, the centriole of the spermatozoon initiated activation and developed into monaster at completion of the second meiosis (Fig. 8-19). The activation of spermatozoon monaster preceded the activation of oocyte aster, and it takes time until the sperm monaster was fully activated. For a while, oocyte centriole remained inactive. Full activation of the sperm monaster followed by activation of oocyte centriole and developed into oocyte aster. In advance of time, activity of the oocyte aster was balanced with the activity to the sperm monaster (Fig. 8-20). In subsequent stage, sperm and oocyte asters approached each other, fused eventually to form nucleus of the zygote (Fig. 8-9, 8-10). The precede activation of the sperm monaster and delayed activation of oocyte aster assume that pronuclei-fusion was initiated by sperm monaster. The fertilization events were summarized in Figure 8-29.

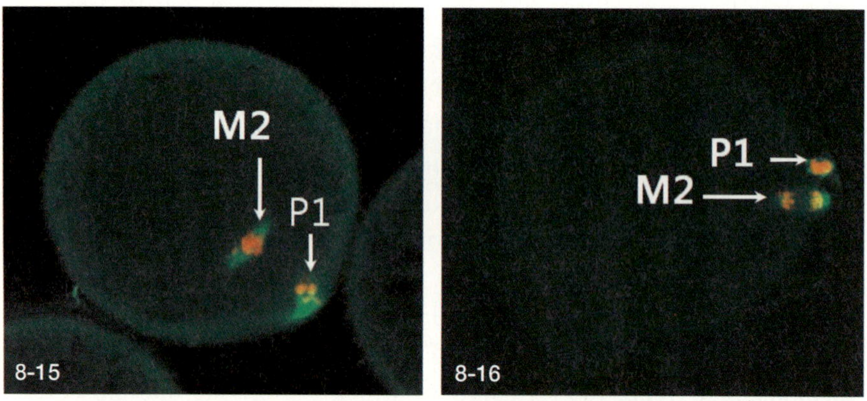

Fig. 8-15 At beginning of the second meiosis, bipolarity of the second meiotic apparatus (M2) is found in middle of ooplam, and arranged parallel to the tangential line of the destined oolemma site. Appearing site of the second meiotic apparatus is differed from the functional site of the second meiotic apparatus near the oolemma (*See* Fig. 14). P1: the first polar body. × 800.

Fig. 8-16 Approaching the destined oolemma site, the second meiotic apparatus (M2) is vertically arranged to the tangential line of the corresponding oolemma site. The second meiotic apparatus is rotated and shifted (rotation-shift) during migration from the middle of ooplasm to the functional site. P1: the first polar body × 800.

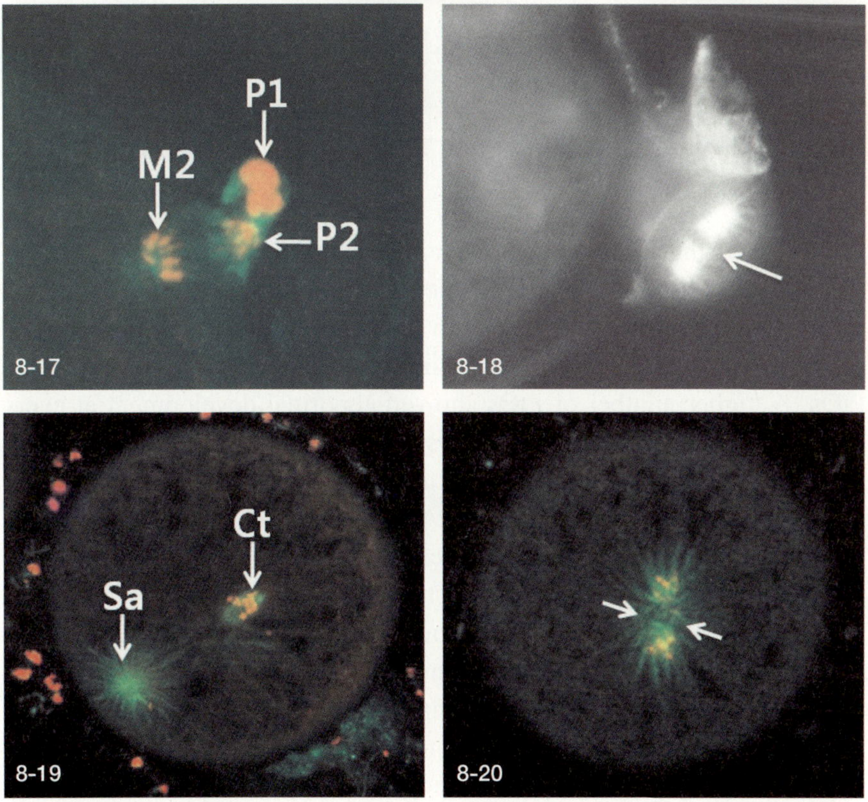

Fig. 8-17 At completion of the second meiosis, the chromosome set of the first polar body is activated and concentrated. The first polar body (P1) is divided into two to form the first and third polar bodies. M2: the second meiotic apparatus, P2: the second polar body. × 800.

Fig. 8-18 Confocal microscopy on the first polar body shows separation of the chromosome set (arrow) during the second meiosis. × 800.

Fig. 8-19 At completion of the second meiosis, centriole of spermatozoon origin produces radial astral rays to form "sperm monaster". The activation of sperm monaster (Sa) precedes the oocyte centriole (Ct). × 800.

Fig. 8-20 The oocyte centriole is delayed in activation. At approaching of the sperm monaster, the oocyte centriole recovers activity to form oocyte monaster equal in activity to sperm monaster. Pronuclei fusion is initiated by sperm monaster. Arrow: Associations of astral arrays of both monasters. × 800.

8-4 Parthenogenetic activation of the oocyte

8-4-1. Immunumofluorescence staining of the parthenogenetic oocyte

To observe the microtubule organizing center (MTOC) or pre-mitotic aster-like structure (PMA) in the parthenogenetic oocyte (PtO), unfertilized oocytes were treated with ammonium acetate-seawater (AASW). Immunofluorescence and CLSM were used to visualize the PMA or MTOC in the PtOs. A higher number of PtOs showed abnormal development at every developmental stage, and many PtOs ceased to develop between the second meiotic cycle and entrance into the cleavage stage. Others frequently exhibited abnormal development at the cleavage stage. However, there were always PtOs in every experiment which developed similar to fertilized oocytes until reached four-cell stage.

PtOs were able to recover their roundness and established fertilization envelopes composed of CG secretions. The two successive meiotic cycles including the formation of a meiotic apparatus, PB extrusions, migration of the meiotic apparatus, and oocyte monaster activation were observed in the PtOs. Compared to the fertilized oocytes, various fertilization events of parthenogenesis were similarly appeared until the first PB was formed. Thereafter, there was a tendency for the parthenogenetic events to be gradually delayed, and development was retarded at the cleavage stage. Differences of development times between the PtOs and fertilized oocytes were summarized in Figure 8-29.

It should be pointed out that the heart-shaped blastomere was a characteristics and prerequisite feature observed at the beginning of normal oocyte cleavage. However, the heart-shaped blastomeres do not distinctly appear in the PtOs. The pattern of *Urechis* oocyte cleavage was equal and holoblastic (Shin *et al.,* 2004), but the PtOs frequently underwent unequal and partial cleavage. As above, development into the blastular stage was observed, but this was rare in the AASW-activated PtOs

8-4-2. The immunofluorescence active particles (IAPs) and pre-mitotic aster-like structure (PMA)

In the most of animal species, the origin of centrioles was known as pre-mitotic aster-like structure (PMA) appeared in unfertilized oocyte, of which development was arrested at a meiotic maturation stage. To observe the origin of oocyte centrioles, *Urechis* oocyte was parthenogenetically activated using ammonium acetate. As described in "material and methods", ammonium acetate was added in sea water, and the unfertilized oocyte was immersed in the ammonium acetate sea water (AASW). Much many of the AASW activated oocytes interrupted developments, however, there were always some oocytes developed to four cell stage. Fig. 8-30 was a comparison of fertilization events expressed by time after sperm addition and parthenogenetically activated oocytes. The events occurred in the fertilized oocyte and AASW activated oocyte were paralleled until extrusion of the second polar body at 90 min.

Germinal vesicle was intact until 2 min in the AASW activated oocyte (Fig. 8-21), as seen in the artificial fertilized oocyte (Fig. 8-11). The Propidium immunofluorescence showed that the diploid chromosomal set was grouped around the nuclear envelope. In reference of literatures of other animal oocyte, appearance of PMA is expected at the eccentric ooplasm between germinal vesicle and oolemma. Although the employed methods revealed cytoskeletal system connecting the germinal vesicle and oolemma and chromosomal set at near the nuclear envelope, the methods resolved no PMA in *Urechis* parthenogenetic oocytes (Fig. 8-21, 9-19).

As seen previously in the fertilized oocytes (Fig. 8-12), GVBD was occurred at expected time of 2 min after the AASW activation (Fig. 8-22). Immediately after the GVBD, nuclear region was distinguished from the marginal ooplasm in presence of diploid chromosomal set, nucleolus and higher material density in the nuclear region. The chromosome condensation and pairing indicated the oocyte progresses zygonema or pachynema stage of meiosis I. At the stage of meiosis, immunofluorescence (anti-tubulin) active particle (IAP) was appeared at border of the nuclear region and oolpasm. Serial observations showed that IAP was transformed into granule like-structure (GLS, 0.4~0.6 μm) equipped with astral rays

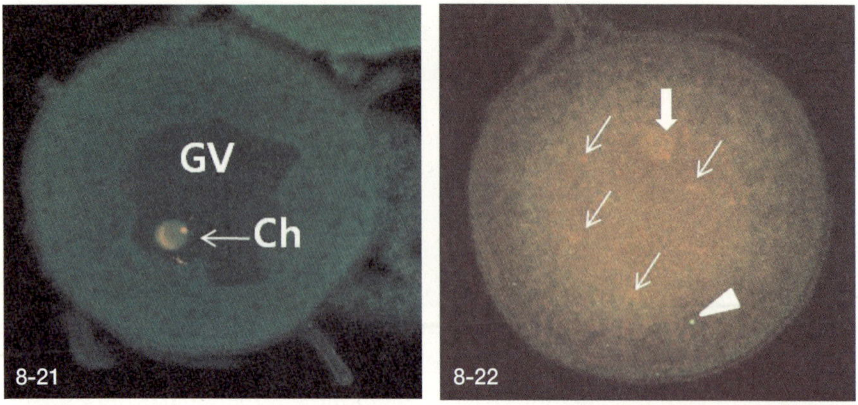

Note All magnification of following parthenogenetic figures is eight hundred times without additional mention.

Fig. 8-21 Parthenogenetically activated *Urechis* oocyte is stained in anti-tubulin-FITC immunofluorescence dye coupled with Propidium iodide. The germinal vesicle including diploid chromosome set (Ch) is intact immediately after parthenogeneic activation. The chromosome is goruped near the nuclear envelope of the germinal vesicle (GV). Before GVBD, original form of oocyte centriole is not to be observed in the expected site of the ooplasm.

Fig. 8-22 At 2-3 min after parthenogenetic activation, the oocyte enters late zygonema or beginning pachynema stage of meiotic prophase I. Chromosome pairs (arrows) are scattered in the nuclear region, at which the germinal vesicle is previously break-downed and nuclear materials are separately concentrated. The IAP (arrow head) is appeared at border of nuclear region and ooplasm opposite to nucleolus. The IAP is analogous to PMA of other animal species. Block arrow: nucleolus.

and the GLS developed further to centrioles of the meiotic oocyte.

The IAP and nuclear region maintain its configuration until diakinesis stage of meiosis I (Fig. 8-25). The site of the IAP appearance was distanced and opposite site from the nucleolus (Fig. 8-22, 8-23, 8-25), differed from the site of PMA of other animal species. The PMA was known to be appeared at eccentric ooplasm between oolemma and germinal vesicle. The IAP was appeared at marginal site of the nuclear region after the occurrence of GVBD in the PtOs. The PMA was found in unfertilized oocyte, but the IAP was not observed in unfertilized or inactivated oocyte. The PMA was differed from IAP in appearance site and time. The difference seemed come from the site or location of IAP in inactivated oocyte. The IAP site was close related with nuclear envelope of the germinal vesicle. It assumed that the IAP was embedded or concealed in the nuclear envelope of the inactivated oocyte.

In addition, at the diplonema and diakinesis stage of meiosis I, the condensing chromosomes make it possible to count number of the chromosome pairs (Fig. 8-23, 8-25). The chromosome pairs were counted as sev-

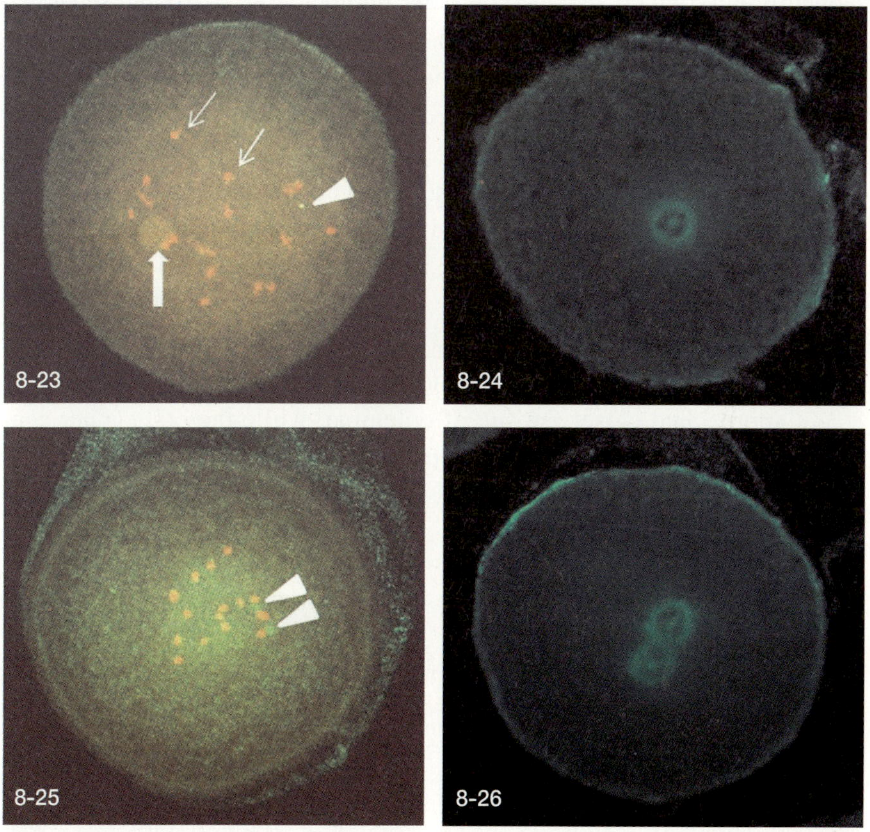

Fig. 8-23 At diplonema stage of meiotic prophase I, the condensed chromosome pairs (arrows) are visualized, counted seventeen (*See* Fig. 25). The IAP (arrow head) is stayed still at border of the nuclear region. Block arrow: nucleolus.

Fig. 8-24 At disappearance of the nuclear region, the IAP produces astral rays, develops into granule like-structure (GLS), which is assumed as centriole of oocyte origin equipped with astral rays.

Fig. 8-25 At diakinesis, the IAP (arrow head) is duplicated, and each of duplication approaches chromosomal pair. As seen in the Fig. 8-23, the chromosome pairs are counted as seventeen, hence the chromosome number of *U. unicinctus* counted out thirty four.

Fig. 8-26 The IAPs develop into GLSs equipped with astral rays, and piled onto one another in middle of the ooplasm. In consideration of the origin, characteristic of duplication and creating of astral rays, it is likely that the GLS is corresponded to centriole of the parthenogenetic activated *Urechis* oocyte (*See* Fig. 8-27, 8-28).

enteen, and in reference of the *Urechis* unfertilized oocyte was pre-meiotic diploid oocyte, the chromosome number of the *U. unicinctus* was evaluated as thirty four.

At around 10 min of AASW activation, the GLS was divided subsequently into two GLSs, piled onto one another (Fig. 8-26) and separated each other as like the IAP (Fig. 8-25). The border of the nuclear region and ooplasm was disappeared, and the nuclear region was no more than distinctly separated. Polarity was created in each of the GLSs, and bipolarity was to be seen in the GLS pair.

At 40 min of AASW activation, the arrangement and unbalanced polarity of the GLS pair was corresponded to the Map approaching oolemma in the fertilized oocyte (Fig. 8-12), and the stage was accorded to end of the first meiosis. The GLS pair or the Map of the parthenogenetic oocyte seemed migrated from the appearing site to near the oolemma. The outer centriole preceded the inner centriole in activation of radial arrays (Fig. 8-27), approached closer the site of oolemma, while the other inner centriole remained behind the outer centriole. The duplication, migration and the creation of asymmetrical polarity of the outer and inner GLS were re-

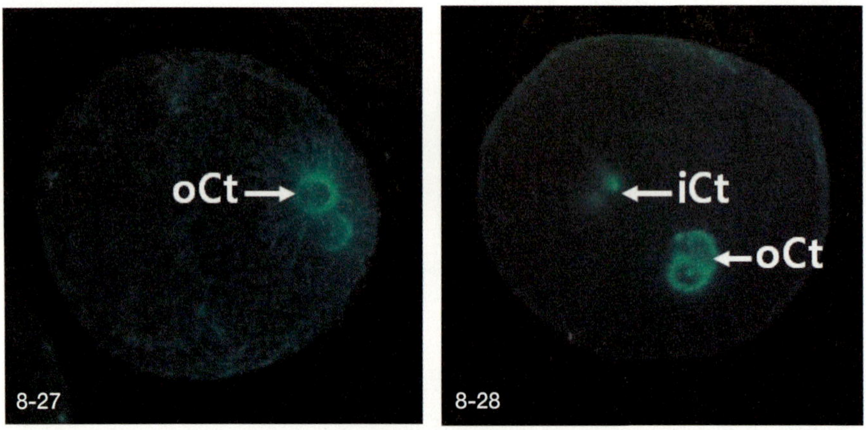

Fig. 8-27 The centriole of higher activity is outer centriole (oCt) of the vertically arranged meiotic apparatus (*See* Fig. 8-13, 8-16). The oCt is the prospective centriole of the first polar body. The stage is accorded to end of first meiosis in reference of the fertilized oocyte.

Fig. 8-28 At beginning of the second meiosis, both centrioles of the oocyte and the first polar body are duplicated. The stage of parthenogenetic oocyte corresponds to beginning of the second meiosis in reference of fertilized oocyte . iCt: inner centrioles, oCt: outer centrioles.

minded of the first meiotic apparatus in the fertilized oocyte.

At around 90 min of AASW activation, both inner and outer centrioles were piled onto one another again showing setting of the duplications (Fig. 8-28). Two pairs of centriole were produced. A pair of centriole was located in deep ooplasm, and the other pair of centriole remained at marginal ooplasm near the oolemma. Comparing with the artificially fertilized oo-

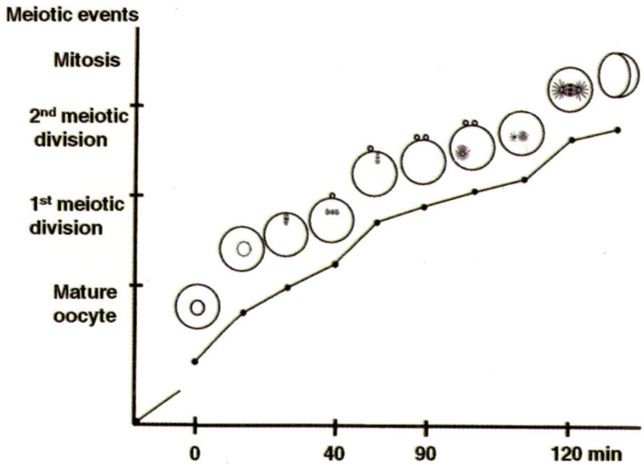

Fig. 8-29 Time table of the fertilized *Urechis* oocyte expressed by meiotic events. At 40 min. and 90 in of fertilization for instance, the oocyte completes the first and second meiosis respectively.

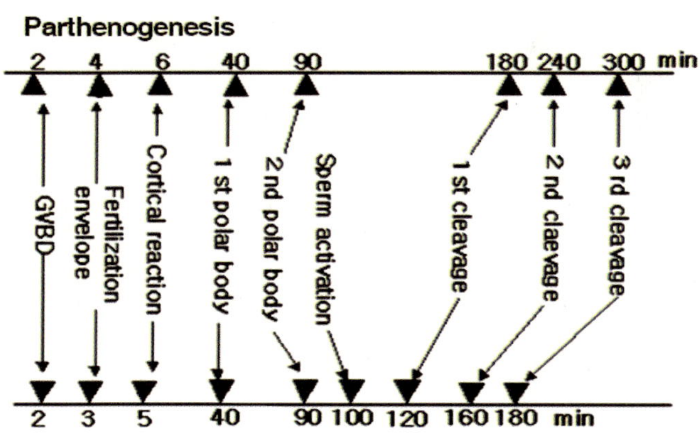

Fig. 8-30 Comparative time-table of the developmental processes between parthenogenetic and fertilized oocytes. The developments of the parthenogenetic oocyte are paralleled with the fertilized oocyte until reached at completion of the second meiosis.

cyte, the duplication of the outer centriole resembled the division of the first polar body and the duplication of the inner centriole was similar to division of the second meiotic apparatus immediately after the completion of the first meiosis.

8-5 The origin of meiotic apparatus

Anti-tubulin-FITC immnofluorescence methods employed in this work resolved IAP in AASW activated *Urechis* oocyte. The IAP was duplicated, developed to granule like-structures (GLSs) piled onto one another producing astral rays prior to the first and second meiosis, respectively. In reference of the GLSs were derived from the IAP, and the GLSs were corresponded to centriole in regard of its polarity, arrangement and migration pattern, it was likely that the IAP was accorded to PMA of other animal species, and the GLSs were identical with the Map of the fertilized *Urechis* oocyte. The observations revealed that the IAP was originated from *U. unicinctus* oocyte. Although the developments of oocytes are activated after fertilization (Gould-Someros, 1975; Gould-Someros & Hollander, 1975, Kwon *et al.*, 2003; Shin *et al.,* 2004), the exogenous origin of the IAPs from the centriole of the fertilized spermatozoon was excluded by the reason of that the fertilized spermatozoon remained inactive at eccentric ooplasm until end of the first meiosis. The centriole of spermatozoon origin was fully activated at the end of the second meiosis, initiated fusion of female and male pronuclei, not related to meiosis of the oocyte.

As the origin of the Map in *Urechis*, the IAP has been directly compared with PMA of other oocytes. PMA appears in unfertilized oocyte and adheres in eccentric ooplasm between oolemma and germinal vesicle (Echinoderm, Schatten, 1994; Miyazaki *et al.,* 2000; Hamaguchi, 2001; mammals, Alexander, 1988; Lopata, *et al.*, 1988). After fertilization, the PMA develops into meiotic apparatus, and the meiosis-arrested oocyte reinitiates developments overcoming the arrested maturation (*Urechis bonellia,* Leutert, 1974; sea cucumber, Miyazaki *et al.*, 2005; *Spiula solid-*

issima, Kuriyama, *et al.,* 1986; *Tubifex hattai,* Ishii & Shimizu, 1997; Echinoid, Schatten, 1994; mollusk, Allen, 1953; Longo & Anderson, 1969; Luetjens & Dorrestein, 1998; amphibians, Alanko, *et al.,* 1994; mammals, Palacios *et al.*, 1993). In those animal species, the meiosis is arrested in presence of cytoplasmic factors, maturation promoting factors (MPF; Masui, 1991; Segata, 1996; Whitaker, 1996; Taieb *et al.*, 1997), maturation inhibiting factor (MIF) and cytostatic factor (CSF, Colas & Dube, 1998; Kishimoto, 1999). As above, much many oocytes in possess of the PMA are arrested at a meiotic process. *Urechis* oocyte is fertilized at germinal vesicle stage (Gould-Somero, 1975; Gould-Somero & Hollander, 1975, Kwon *et al.*, 2003; Shin *et al.,* 2004), and once being fertilized, the GVBD, meiotic cycles and pronuclear fusion are continuously progressed without arrest or interruption. For the reason, it was referred that the IAP was analogous to PMA in regard of the function.

The presence of PMA is reported in one of *Urechis* species, *Urechis bonellia* (Leutert, 1974). However, there were some differences between IAP and PMA in reference of the reports on the PMA. First, the IAP was observed in *Urechis* oocyte which was activated at GV stage. The PMA is known to be appeared in kinds of oocyte of which maturation is arrested at a meiotic cycle, and fertilized at a stage between prophase I and metaphase of the second meiosis. The oocyte type of IAP appearance seemed differed from the oocyte type of PMA appearance.

Second, the site and time of IAP appearance was differed from the PMA site and time. The PMA appears in eccentric ooplasm of unfertilized oocyte, while the IAP was not observed in unfertilized and inactivated oocyte, but observed at time of GVBD in the AASW activated oocyte. Direct after the GVBD, the IAP was arisen in the nuclear region bordered on ooplasm. The nuclear region was site of nuclear materials being concentrated in the region of the germinal vesicle was previously break-downed. The IAP seemed closely associated or concealed in nuclear envelope of the germinal vesicle. For the reason, the IAP was appeared simultaneously with occurrence of GVBD.

Third, it is known that the PMA is assigned to proper function. The PMA is connected with oocyte cytoskeletal system plays role in relocation and protection of germinal vesicle (Hamaguchi, 2001). PMA protects the position of the germinal vesicle from extreme maldistribution, and promotes the relocation of the germinal vesicle to the prospective site of PB extrusion. The relocation of germinal vesicle is established by PMA or polymerization and depolymerization of cytoskeletal actin microfilaments (Lutz *et al.*, 1988; Gard *et al.*, 1995; McGraill & Hays, 1997; Skop & White, 1998) under influence of Ca^{++} concentration (Dube & Gurrier, 1982; Brundage *et al.*, 1991), pH, and a set of actin binding proteins (Finkel & Wolf, 1981; Verkovsky & Borisy, 1993). A proper function of IAP is not known. As above, the *Urechis* IAP was concealed in the nuclear envelope of unfertilized oocyte. Although the employed immunofluorescence methods revealed cytoskeletal system connecting germinal vesicle to all sites of oolemma, the presence of IAP was not resolved in the unfertilized oocyte (Fig. 9-19). *Urechis* oocyte is isolecithal; germinal vesicle is located at equal distances from all sites of the oolemma, arranged always at the middle position of the oocyte (Shin *et al.*, 2004). Role of IAP in the unfertilized oocyte is not known. Cytoskeletal system was revealed in unfertilized *Urechis* oocyte (Fig. 9-19), but connection of the cytoskeletal system with the IAP was not observed in *Urechis* oocyte. Astral rays were observed on GLS after GVBD.

In addition, PMA appears as above in type of oocyte fertilized at a stage of meiotic arrest. The extrication from the maturation arrest and reinitiation of oocyte developments necessitate functional cytoplasmic factors after the occurrence of fertilization. It is suspected whether the cytoplasmic factors are realized in *Urechis* oocyte, since the IAP was concealed in nuclear envelope, appeared after GVBD, and developments of IAP to centriole of the Maps were continuously progressed without interruption or arrested developmental stages (Kwon, *et al.*, 2003; Shin *et al.*, 2005a; Shin *et al.*, 2005b).

8-5-1. Rotation-shift of Urechis bipolar spindle fibers

In *Urechis* oocyte, the Map was derived from the IAP appeared in border of nuclear region and ooplasm, where the germinal vesicle was break-downed. At end of first and second meiosis, the Maps were arranged vertical to tangential line of the oolemma, and the active outer centriole was closely associated with site of the destined oolemma. Since polar body

(PB) was extruded at the oolemma site remaining inactive inner centriole behind, the Maps accomplished it function at the oolemma site. In summary, the Maps migrated from the appearing site in the middle of oocyte to the functional site near the oolemma. Polarity of the Map was changed asymmetrical during the migration. In consequence, the active outer centriole with a set of chromosomes was selectively eliminated, while inactive inner centriole was remained in the oocyte, as is described in starfish (Saiki & Hamaguchi, 1997). The event was repeated in the second meiosis. Finally, one inactive centriole was remained in the oocyte; others were eliminated by mean of PB formation being extruded out of the oocyte.

The differences of appearing site from the functional sites were compensated by migration of the Maps to the destined site of oolemma. During the migration, the *Urechis* Maps showed characteristic pattern in arrangement of the Map bipolarity. At appearing in the middle ooplasm, the second bipolar Maps were arranged parallel to the tangential line of the oolemma site, the prospective PB site. In front of the PB site at the end of migration, the Map was arranged vertical to the oolemma in both fertilized and parthenogenetic oocytes (PtOs) of *Urechis*. As consequence, the bipolar Map in *Urechis* oocyte was rotated and shifted during the migration. The "rotation-shift" of Map is reported previously in starfish, compensates the site differences of the Maps (Hamaguchi, 2001). In *U. unicinctus*, the change in arrangement of the first map was not exactly observed. However, the processes of the first meiosis were identical to the second meiosis including the vertical arrangement and creation of asymmetrical polarity in the first Map, and it is considered that the first Map was arranged parallel to the oolemma. At beginning of the migration, bipolarities of the Maps were created in both fertilized and PtOs, maintained for a while during the migration. Approaching the PB site, the bipolarity was changed. The outer pole became highly active, while inner pole is relatively weakened. The changes of bipolarities are similar to starfish spindle fiber (Schatten, 1994; Miyazaki *et al.,* 2000; Tamura & Nemoto, 2001), but distinguished from the monopolar spindle fibers in the annelid *Tubifex Hattai* (Ishii & Shimizu, 1997). The creation of polarity and migration of the Maps are caused by similar signal transduction system of the Maps and oolemma (Lutz *et al.*, 1988; Miyazaki, *et al.,* 2000).

With exception of mice, the process creating different polarity and the selective activation of the centriole is reported in various animals (Schatten *et al.*, 1985; Schatten, 1994), modeled as a regular pattern of meiosis. It was considered that the parthenogenetic and artificially fertilized *U. unicinctus* oocyte was made no exception, conformed to the modeled pattern. The pattern is the consequence of one non-replicative centriole remains finally in the ooplasm, and the other replicative centrioles are extruded to produce PBs. On the cause of selective activation, previous studies have suggested activation of enzymes such as MAP kinase (Gould-Somero & Stephano, 1999), inhibition of oocyte centrioles by the sperm centriole (Sluder *et al.*, 1989; Sluder *et al.*, 1985), and programmed regulation of oocyte development (Sluder *et al.*, 1993; Schatten 1994). This study has not followed in detail the selective activation and reduction of *Urechis* meiotic centrioles, but the inhibition by the sperm centriole can be excluded from the cause of the selective activation. The fertilized spermatozoon remained inaction at an ooplasm, its activation began at end of the first meiosis and full activation was observed at completion of the second meiosis. Before the centriole of spermatozoon origin was fully activated, two meiotic cycles have already been completed in *U. unicinctus* oocyte.

Based on the patterned selective activation of the meiotic centrioles, the term "meiosis" is commonly used to define inactivation processes of centriole for production of haploid oocyte. A pair of oocyte centrioles is replicated to produce two centriole pairs which are equally distributed into each of the three PBs and oocyte. The replicative and divisional centrioles are continuously eliminated through successive reduction resulting in half of the four centrioles (Saiki & Hamaguchi, 1998; Tamura & Nemoto, 2001). In parthenogenetic and fertilized oocyte of *U. unicinctus*, the second and third PB is inactivated shortly after their formations, while centrioles of the oocyte and first PB are reactivated. Both centrioles derived from the first polar body and one centrioles derived from the oocyte were subsequently eliminated. Although *Urechis* oocyte is fertilized at germinal vesicle stage, the meioses showed typical pattern of successive centriole reduction that eliminated three of the four centrioles. In consequence, the non-replicative monaster remained in the oocyte.

Differed from centriole of the second PB, it is known in patterned meiosis that the centriole of the first PB is divided producing two PBs in the patterned meiosis. The centriole of the first PB differs in divisional capacity from the centriole of the second PB. During parthenogenetic activation of the starfish, the first PB is transplanted into the oocyte of which centrioles are previously inactivated. The recipient oocyte is divided to extrude PB and once when the PB has been extruded, the oocyte is not able to divide further (Nemoto *et al.*, 1994; Saiki & Hamaguchi, 1998). For this reason, the oocyte is not able to divide further when the third PB has been extruded. As the consequence, one inactive centriole is remained in the oocyte and others are eliminated extruding out of the oocyte. On the other hand, when PB is prevented from extruding, the oocytes are able to divide further (Sluder, 1989; Sluder *et al.,* 1993; Nemoto *et al*., 1994; Miwa & Nemoto, 2001). In *U. unicinctus*, the remained centriole in the oocyte was inactive at completion of the second meiosis. Activation of sperm monaster was followed by activation of the centriole remained in the oocyte; nuclear fusion seemed initiated by the sperm monaster.

Summary

To observe the origin of meiotic apparatus (Map), anti-tubulin-FITC and Propodium iodide are applied on unfertilized, fertilized and parthenogenetic oocyte of *Urechis unicinctus.* The immunofluorescence active particle (IAP) is observed as the origin of centrioles in *Urechis* meiotic cell. The IAP is compared with pre-mitotic aster- like structure (PMA) which is known as the origin of centrioles in other animal species. Similar to PMA, the IAP is duplicated and developed into granule-like structure (GLS) with astral rays and the centrioles of meiotic apparatus (Map). However, the IAP is differed from the PMA in various points of view. The IAP is not found in unfertilized on parthenogenetic inactivated oocyte, but observed in parthenogenetic activated oocyte after occurrence of GVBD. The appearing site and time of IAP are differed from the appearing site and time of PMA. The reason is assumed that the IAP is concealed in nuclear envelope of germinal vesicle. Since connections between IAP and cytoskeletal system are not observed in unfertilized or parthenogenetic inactivated oocyte, it is suspected that the IAP is employed in positional regulation of the germinal vesicle in the isolecithal oocyte. The IAP is analogous to the PMA.

Meiotic process of the parthenogenetic oocyte is compared with the meiosis of fertilized oocyte. The appearing site of the Map is differed from the functional site. Regular pattern of meiotic apparatus including rotation-shift of the Map during migration, creation of differential polarity of the centrioles, continuous eliminations of active centrioles during meiotic processes and remain of an inactive and nonreplicative centrioles in the oocyte are observed. *U. unicinctus* oocyte is conformed to the modeled pattern of meiosis, although the oocyte is subordinated to the oocyte type which is fertilized at germinal vesicle stage.

REFERENCES

Alanko T, J Tienari, E Lehtone, O Saksela: γ-Tubulin is asymmetrically distributed in the cortex of Xenopus oocytes. Dev Biol, 161: 131-140, 1994.

Alexander AJ, The fertilization of human oocytes at different stage of meiotic maturation. "*In vitro* Fertilization". Howard W Jones Jr.ed. Ann NY Acad sci 541, p324, 1988.

Allen RD: Fertilization and artificial activation in the egg of surf calm *Spisula solidissima*. Biol Bull 105: 213-239, 1953.

Bestor TH, G Schatten: Anti-tubulin immunofluorescence microscopy of microtubules present during the pronuclear movements of sea urchin fertilization. Dev. Biol, 88: 80-91, 1981.

Brachet J: Oogenesis In "Developmental biology" 3rd ed, Browder LW ed, Saunder coll. Pub., pp 55-116, 1991.

Brundage RA, KA Forgaty, RA Tuft: Calcium gradients underlying polymerization and chemotaxis of eosinophile. Science, 254: 703-706, 1991.

Colas P, F Dube: Meiotic maturation in molluse oocytes. Semin Cell Dev Biol, 9: 539-548, 1998.

Dube F, Guerrier P: Activation of *Barnea candida* (Mollusca, Pelecypoda) oocytes by sperm of KCl, but not by NH_4Cl, requires a calcium influx. Dev Biol 92: 408-417, 1982.

Finkel T, Wolf DP: Membrane potential, pH and the activation of surf calm oocytes. Gamete Res 3: 299-304, 1981.

Gard DL, B Cha, AD Roder: F-actin is required for spindle anchoring and rotation in Xenopus oocytes: A reexamination of the effects of cytochalasin B on oocyte maturation. Zygote, 3: 17-26, 1995.

Gould-Somero M: Echiura. "In Reproduction of Marine Invertebrates". A C Giese & J S Pearse ed, 3: 277-311, Academic Press, Inc., New York, 1975.

Gould-Somero M, L Holland: Fine structural investigation of the insemination response in *Urechis Caupo*. Dev Biol 46: 358-369, 1975.

Gould-Somero M, JL Stephano: MAP kinase, meiosis, and sperm centrosome suppression in *Urechis caupo*. Dev Biol. 216: 348-358, 1999.

Hamaguchi Y: Displacement of the mitotic apparatus which induces ectopic polar body formation or parthenogenetic cleavage in the star fish oocytes. Dev Biol, 364:364-375, 2001.

Ishii R, T Schimizu: Equalization of unequal first cleavage in the *Tubifex* egg by introduction of an additional centrosome : Implications for the absence of cortical mechanisms for mitotic spindle asymmetry. Dev Biol, 189: 49-56, 1997.

Kishimoto T: Activation of MPF at meiosis reinitiation in starfish oocytes. Dev Biol, 214: 1-8, 1999.

Kuriyama R, GG Borisy, Y Masui: Microtubule cycles in oocytes of the Surf clam, *Spisula solidissima*: An immunofluorescence study. Dev Biol, 114: 151-160, 1986.

Kwon HJ, JW Jeong, WJ Kim, KS Shin: Immunofluorescence and electron microscopic study on the artificial insemination and rotation-shift behaviors of the bipolar spindle fiber in *U. unicinctus* egg. Korean J Electron Microscopy, 33(2) 105-116, 2003.

Leutert TR: Zur geschlechtsbestimung und gametogenese von Bonellia vividis. J Embryol Exp Morphol, 32: 169-193, 1974.

Longo FJ, E Anderson: Cytological aspects of fertilization in the lamellian branch *Mytilus edulis* I. Polar body formation and development of the female pronucleus. J Exp Zool 172: 69-96, 1969.

Lopata A, PC Leung: The fertilization of human oocytes at different stages of meiotic maturation. In "*In vitro Fertilization*". Howard W Jones Jr. ed. Ann NY Acad Sci, 541: pp324, 1988.

Luetjens CM, AWC Dorrestein: Dynamic changes of the microtubule system corresponding to the unequal and spiral cleavage modes in the embryo of the zebra mussel, *Dreissena polymorpha* (Mollusca, Bivalva. Zygote, 6:239-248, 1998.

Lutz DA, Y Hamaguchi, S Inous: Micromanipulation studies of the asymmetric positioning of the maturation spindle in Chaetopterus sp. oocytes. I. Anchorage of the spindle to the cortex and migration of a displaced spindle. Cell Motil Cytoskeleton, 11: 83-96, 1988.

Masui Y: The role of "cytostatic factor (CSF)" in the control of oocyte cell cycles : a summary of 20 years of study. Dev Growth Differ, 33: 543-551, 1991.

McGrail M, TS Hays: The microtubule motor cytoplasmic dynein is required for spindle orientation during germline cell division and oocyte differentiation in *Drosophila*. Development, 124: 2409-2419, 1997.

Miwa T, SI Nemoto: Reproductive maternal centrosomes are cast off into polar body during maturation division in starfish oocytes. Exp Cell Res, 269 : 130-139, 2001.

Miyazaki A, E Kamitsubo, SI Nemoto: Premeiotic aster as a device to anchor the germinal vesicle to the cell surface of the presumptive animal pole in starfish oocytes. Dev Biol, 218: 161-171, 2000.

Miyazaki A, KH Kato, SI Nemoto: Role of microtubules and centrosomes in the eccentric relocation of the germinal vesicle upon meiosis reinitation on sea-cucumber oocytes. Dev Biol, 280: 237-247, 2005.

Nemoto WS, C Saitoh, SI Nemoto: Artificial parthenogenesis in starfish egg: Behavior of nuclei and chromosome resulting in tetraploidy of parthenogenotes produced by the suppression of polar body extrusion. Dev Biol, 163:293-301, 1994

Palacios MJ, C Harish, HC Joshi: γ-Tubulin reorganization during mouse fertilization and early development. J Cell Sci, 104: 383-389, 1993.

Saiki T, Y Hamaguchi: Division of polar bodies induced by their enlargement in the starfish *Asterina pectinifera*. Exp cell Res, 237: 142-148, 1997.

Saiki T, Y Hamaguchi: Aster-forming abilities of the egg, polar body, and sperm centrosome in early starfish development. Dev Biol, 203: 62-74, 1998.

Schatten G: The centrosome and its mode of inheritance: The reduction of the centrosome during gametogenesis and its restoration during fertilization. Dev Biol, 165: 299-335, 1994.

Schatten G, C Simerly, H Schatten: Microtubule configurations during fertilization, mitosis and early development in the mouse and the requirement for egg microtubule-mediated motility during mammalian fertilization. Proc Natl Acad USA, 82: 4152-4156, 1985.

Schultz AW : Action of hormones on germinal vesicle breakdown in frog(*Rana pipiens*) oocytes.

J Exp Zool, 166 : 347-354, 1967.

Segata N: Meiotic metaphase arrest in animals oocytes : its mechanisms and biological significance. Trends Cell Biol, 6 : 22-28, 1996.

Shin KS, DH Lee, TY Ko: Cleavage pattern of *Urechis unicinctus* egg in vitoro fertilization. Korean J Electron Microscopy, 34(1): 71-81, 2004.

Shin KS, HJ Kim, HJ Kwon, WJ Kim: Cytochemical localization of nuclear actin of sperm and spermatids in *Urechis unicinctus*. Integ, Biosci 9: 65-73, 2005a.

Shin KS, HJ Kwon, WJ Kim: Attribution of cortical granules to formation of fertilization envelope and polyspermy block in *Urechis unicinctus.* Integrative Biosciences, 9: 57-64, 2005b.

Skop AR, JG White: The dynactin complexes are required for cleavage plane specification in early *Caenorhabditis elegans* embryos. Curr Biol, 8: 1110-1116, 1998.

Sluder G, CL Rider: Experimental separation of pronuclei in fertilized sea urchin eggs chromosomes do not organize a spindle in the absence of centrosome. J Cell Biol, 100: 897-903, 1985.

Sluder G, FJ Miller, K Lewis, ED Davison, CL Rider: Centrosome inheritance in starfish zygotes: selective loss of the maternal centrosome after fertilization. Dev Biol, 131: 567-579, 1989.

Sluder G, FJ Miller, K Lewis: Centrosome inheritance in starfish zygotes Ⅱ: selective sppression of the maternal centrosome during meiosis. Dev Biol, 155: 58-67, 1993.

Taieb F, C Thibie, C Jessus: On cyclins, oocytes, and egg. Mol Reprod Dev, 48: 397- 411, 1997.

Tamura M, S Nemoto: Reproductive maternal centrosomes are cast off into polar bodies during maturation division in starfish oocytes. Exp Cell Res, 269: 130-139, 2001.

Uetake Y, KH Kato, WN Setsuko, SN Nemoto: Nonequivalence of maternal trosomes/centrioles in the starfish oocytes: selective casting-off of reproductive centrioles into polar bodies. Dev Biol, 247; 149-164, 2002.

Verkovsky AB, GG Borisy: Nonsarcomeric mode of myosin II organization in the fibroblast lamellum. J Cell Biol, 123: 637-652, 1993.

Whitaker M: Control of meiotic arrest. Rev Reprod, 1: 127-135, 1996.

Chapter 09

Cleavage pattern of *U. unicicntus* oocyte after artificial fertilization

Chapter 09 Cleavage pattern of *U. unicicntus* oocyte after artificial fertilization

9-1 Introduction

Urechis has been used as a laboratory animal in various fields including cell biology (Davis, 1986), biochemistry (Tsutomu *et al.,* 1983), cytochemistry (Muneoka & Kamura, 1982), developmental biology (Engstrom, 1971; Newby, 1940, 1986; Gould-Somero & Hollander, 1975) and phylogenetic studies (Halanych, 1995; Nielsn, 1996) although there are some difficulties in studying the worm. One of these difficulties is the limited range of worm habitats. Three species, *U. unicinctus* (von Drasche), *U. chinensis* (M. Mueller), and *U. caupo* (Fischer), are the dominant species along the Korean, Japanese, Chinese, and Californian coasts (Sato, 1939). The other difficulty is that there is no consistent classification system for these worms as mentioned in the previous chapter. The worm has been classified as an echinoderm (Sato, 1939), but the unsegmented external features and *Hox* cluster and 18S rRNA resemble that of annelids (McHugh, 1997). Furthermore, internal physiologies and structures of the worm including the reproductive system are similar to those of mollusks (Newby, 1940).

Even though embryology and oocyte cleavage patterns may be used to help classify the worm, it is difficult to find literature on worm embryology besides a previous report by Sato (1939). Nevertheless, embryonic developmental patterns are often used as a criterion to classify animals, especially marine invertebrates. The gametes of *U. unicinctus* are artificially fertilized to understand embryology of the worm (Shin *et al.,* 2004). Under laboratory conditions, the zygotes undergo the entire processes of cleavages and stage distributions including trocophore hatching and production of crawling larva. The heart-shaped blastomeres appear prior to the first cleavage, as is reported in sea urchin (He & Dembo, 1997). After the formation of asymmetric heat shaped blastomere, the cleavage restores the symmetric pattern of the blastomeres at least until the completion of the fifth cleavage. At end of the second cleavage, it is observed that two blastomeres are arranged at upper position on the other two blastomeres. At end of the third cleavage, the upper four blastomeres are arranged obliquely on the lower four blastomeres.

The cleaving embryos are incorporated with anti-tubulin-FITC conjugate. The immunofluorescence staining show the mitotic apparatuses (MtAs) in blastomeres of the embryos (Shin *et al.,* 2004). Under a confocal laser scanning microscope (CLSM), the MtAs appear obliquely tilted at every cleavage stage. Tilting of the MtAs are obviously to be observed especially in proceeding of third cleavage, ensures that the *Urechis* zygote cleaves in a spiral pattern.

Spiral cleavage pattern is known in *Spiralian* of some mollusks (Reverberi, 1971), annelids (Freeman & Lundelius, 1982), flatworms (Reververi, 1971), and nemertine worms (Gilbert, 1988). However, the most of *Spiralian* cleaves unequally producing macromeres, mesomeres and micromeres. *Urechis* blastomeres followed in principle the spiral cleavage pattern, even though the cleavage of *Urechis* is equal and holoblastic. The combination of spiral cleavage producing equal blastomeres seemed particular and unique among animal species, cannot be usually found in the previous literature (Shin *et al.,* 2004).

9-2 Cleavage

The oocytes are isolated from the storage sac during the higher spawning season. Cytoplasmic indentations of the oocytes disappear in natural sea water (NSW; 16~18℃, 20‰) (Fig. 9-1, 9-2), which is a kind of brackish water overflowing the worm habitats during the breeding season as previously described. The round oocytes measures 130~150 μm in diameter, germinal vesicle and nucleolus are to be observed in the yellowish semi-transparent ooplasm. The animal and vegetal poles are hardly distinguished, although germinal vesicle (50 μm) is located at slight eccentric position of the ooplasm.

The fertilization events are previously described. Briefly, the fertilization cone rises from the oocytes 2 min after fertilization. Germinal vesicle break-down (GVBD) occurs at around 5 min (Fig. 9-3), and the fertilization envelope is 14~16 μm thick 15 min after fertilization (Fig. 9-4). The first and second polar bodies extrude at around 40 min (Fig. 9-4) and 90

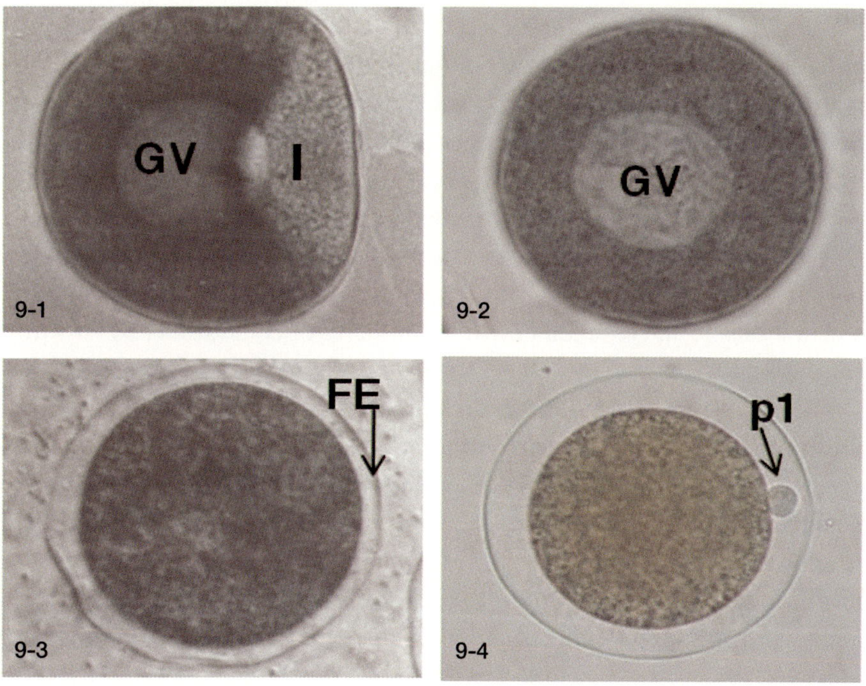

Fig. 9-1 Mature oocyte prepared from female storage sac shows cytoplasmic indentation (I). GV: germinal vesicle. × 400.

Fig.9-2 The oocyte restores round shape after spawning into brackish water, and fertilized at germinal vesicle (GV) stage. × 400.

Fig. 9-3 Germinal vesicle is break-downed and fertilization envelope (FE) is raised after fertilization. Numerous surplus sperm are visible around the fertilized oocyte. × 400.

Fig. 9-4 The first polar body (P1) of the living oocyte is extruded into fertilization envelope at completion of the first meiosis. × 300.

min (Fig. 9-5), respectively. The first polar body produces two polar bodies and three polar bodies are observed at end of the second meiosis (Fig. 9-5). The female and male pronuclei are fused by initiation of the monaster originating from the spermatozoon(Fig. 8-19, 8-20). In *Urechis* oocytes, fertilization occurs at germinal vesicle stage, all fertilization and meiotic events ensure after the entrance of the spermztozoon into the oocyte cortex. The spermatozoon is inactively located in the ooplasm of the oocyte until the zygote fulfills the first meiosis.

9-2-1. First cleavage

During the first cleavage, *U. unicinctus* zygotes form a heart-shaped blastomere in whole configuration (Fig. 9-6). Beginning from the animal pole, the first cleavage furrow continues to invaginate along the median plan of the blastomere. For a while, the vegetal pole is slightly projected so that the unilateral and asymmetrical shape of the whole blastomere reminded in the shape of a heart. Bilateral symmetry is restored as soon as the counter-cleavage furrow began to cleave the zygote from the vegetal pole (Fig. 9-7). The bilateral cleavage furrows met one another at the central portion of the blastomeres. The first cleavage is meridional, symmetric, and holoblastic and resulted in two blastomeres of equal size (120 × 85 μm) (Fig. 9-8).

The heart-shaped blastomeres are prepared for EM observation to search for the contractile ring at the front of the cleavage furrow. Several EM images of the cleavage furrow are taken to construct a mosaic and redrawn to generate an entire view of the cleavage furrow (Fig. 9-25, 9-26). At the leading edge of the cleavage furrow, the membrane components, cellular protuberances, and thickening of the furrow are observed. These features are similar to the reported structural characteristics of the contractile ring observed under EM (Szollosi, 1970; Asnes &Schroeder, 1979; Bestor & Schatten, 1981; Kitajima & Hamaguchi, 2005). The cleavage furrow is composed of narrowed parallel arrays of the cell membrane separated from each other by gap of 8 nm in width. Extension of the cleavage furrow is not continuous, but interrupted at various sites at which dense bodies (7~10 nm) are frequently found. These bodies are observed mainly ahead of the prospective furrow including the site where both cleavage furrows met. The dense bodies tended to be concentrated at the appeared site, but dispersed lately once the furrow passed through the site of the dense body. The fine particles derived from the dispersed dense body merged primarily on the cell membrane of the blastomeres.

The blastomeres are enclosed by outer extracellular envelope derived from the fertilization envelope of the zygote. The heart-shaped zygote in the extracellular envelope is significantly shortened by 25% in vertical length (length between the animal and vegetal pole) and widened horizontally by about 30%. The animal pole site of the extracellular envelope is re-

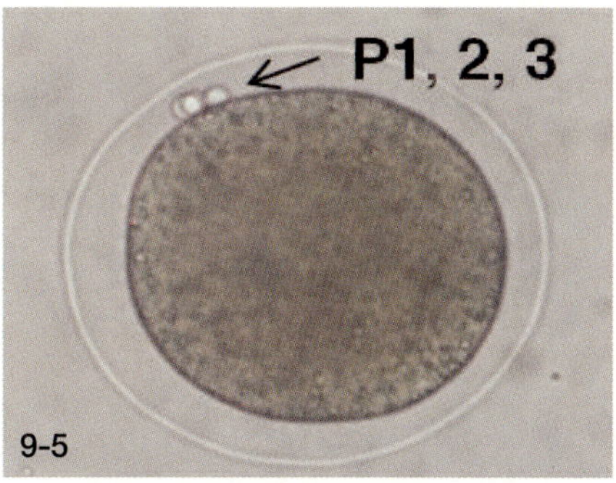

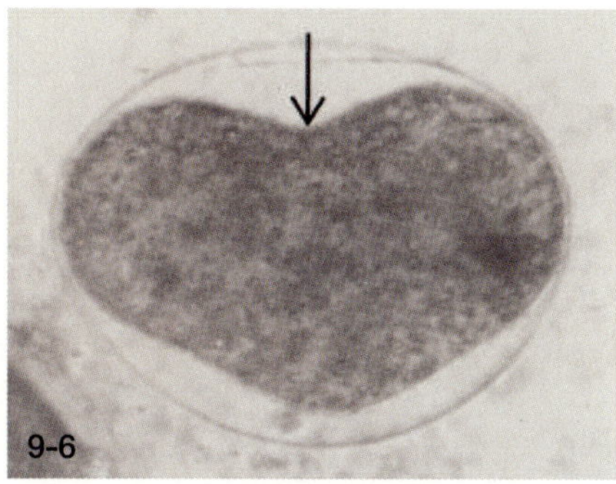

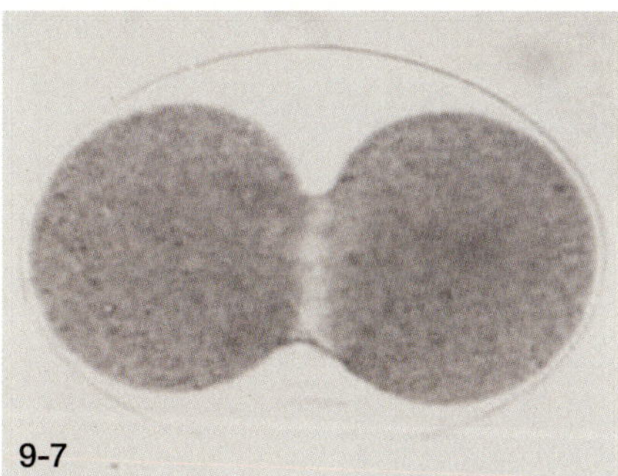

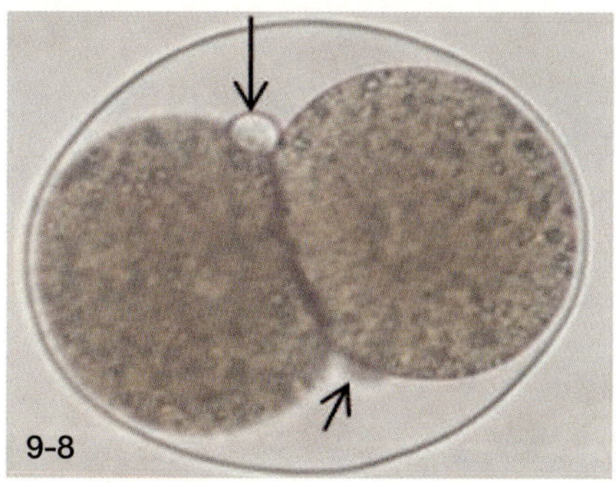

Fig. 9-5 At completion of the second meiosis, three polar bodies of the living oocyte are extruded into fertilization envelope. P1, 2, 3: the first, second and third polar body. × 400.

Fig. 9-6 In proceeding of the first cleavage, heart-shaped blastomere is produced, bilateral symmetry is receded from the developing embryo. Arrow: invagination of cleavage furrow from the animal pole. × 400.

Fig. 9-7 At a late stage of the first cleavage, the embryo is balanced with two prospective blastomeres of equal sizes, recovers bilateral symmetry. × 400.

Fig. 9-8 Polar bodies are scattered in the fertilization envelope at the end of of the first meiosis. Arrows: polar bodies. × 400.

duced to form slightly concave structure, similar to the inside of the envelope under negative pressure. Anti-tubulin immunofluorescence staining show that the MtA of the first cleavage is dextrally tilted with respect to the animal-vegetal axis (Fig. 9-20). After completion of the first cleavage, one upper blastomere is slightly displaced obliquely to the right side of the lower blastomere (Fig. 9-8).

9-2-2. Second and third cleavage

The second cleavage is meridonal, symmetric, and holoblastic, resulting

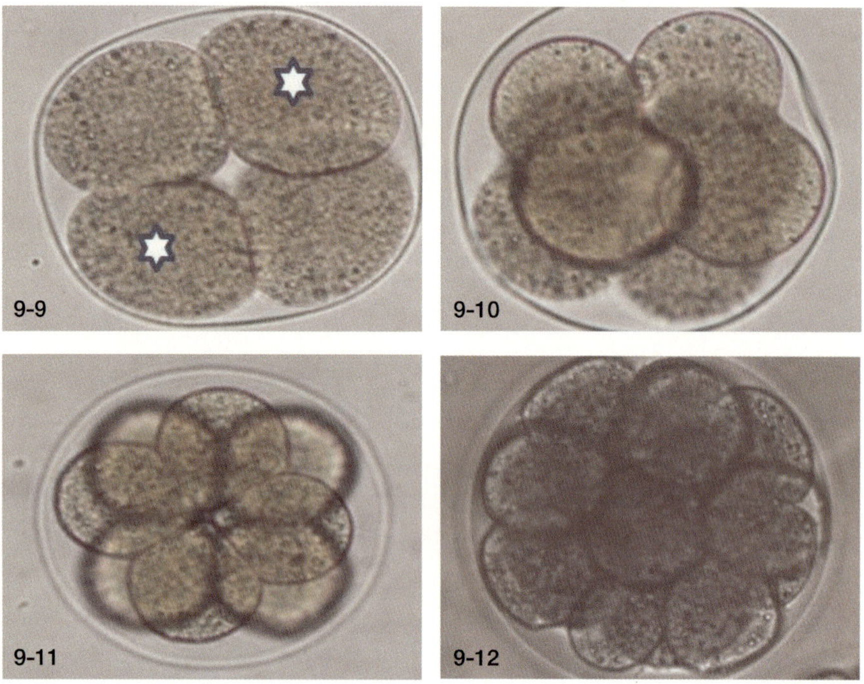

Fig. 9-9 After the second cleavage, two upper blastomere (asterisks) are oblique displaced on the other two blastomeres. The displacement of blastomeres is caused by sinistral tilting of the second bipolar spindle fibers, as seen in fig. 9-21 and 9-22. × 400.

Fig. 9-10 The upper four blastomeres are displaced to right site to the lower four blastomeres after the third equatorial cleavage. × 400.

Fig. 9-11 The displacement is continued until the upper four blastomeres are arranged obliquely to the lower four blastomeres. × 400.

Fig. 9-12 The fourth cleavage is holoblastic cleavage producing equal sized sixteen blastomeres. × 400.

in four blastomeres of equal size (75 μm in diameter; Fig. 9-9). MtAs of the second cleavage are vertically parallel to each other and to the animal-vegetal axis. Over time, the MtAs obliquely tilted counterclockwise, and MtAs being tilted opposite to one another are observed during the second cleav-

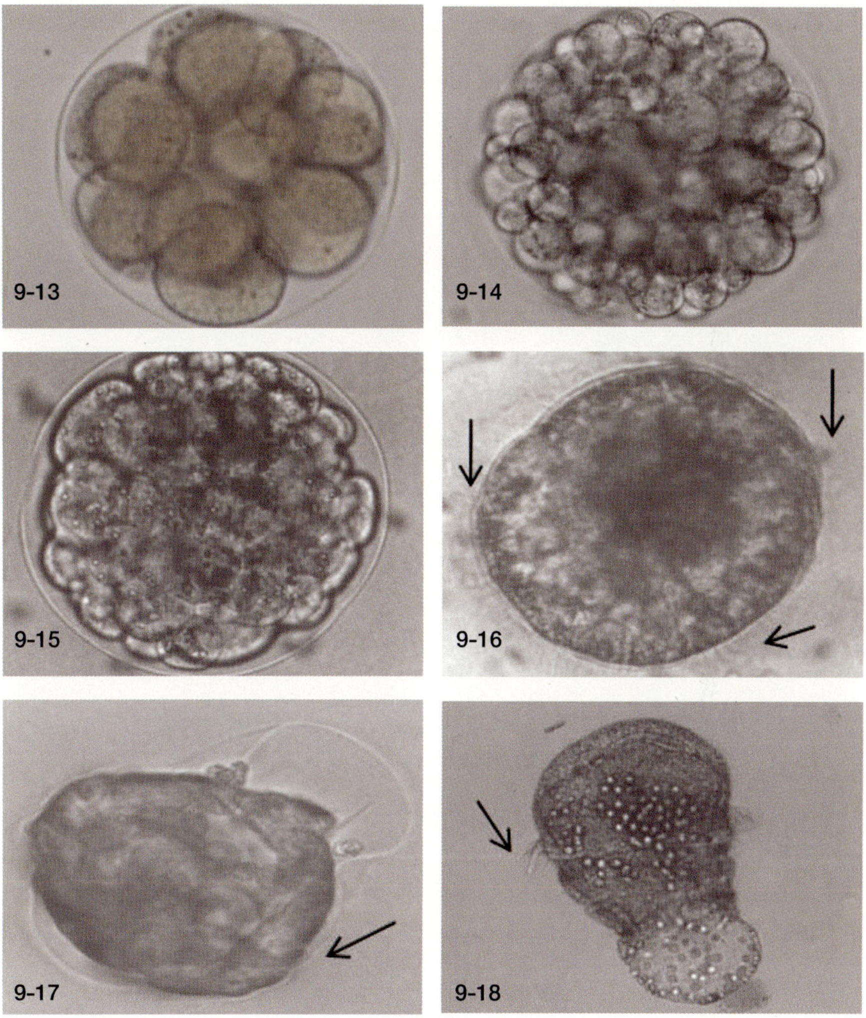

Fig. 9-13 Living embryo going from sixteen cell stage to subsequent stage. X 400.
Fig. 9-14, 9-15 Blastular stage. × 400.
Fig. 9-16, 9-17 Ciliated trocophore. Arrows: cilia. × 400.
Fig. 9-18 Hatched worm with cilia (C). Arrow: cilia. × 400.

age (Fig. 9-21, 9-22). After the second cleavage is completed, the two upper blastomeres of the four cells are arranged obliquely to the left side of the lower blastomere (Fig. 9-9) since two daughter cells are slightly higher than the others positions with respect to the animal-vegetal axis.

The third cleavage is equatorial, symmetric, and holoblastic, and resulted in eight blastomeres 60 μm in size (Fig. 9-10). Initially, the MtAs of the diagonally-positioned blastomeres are parallel to each other and verti-

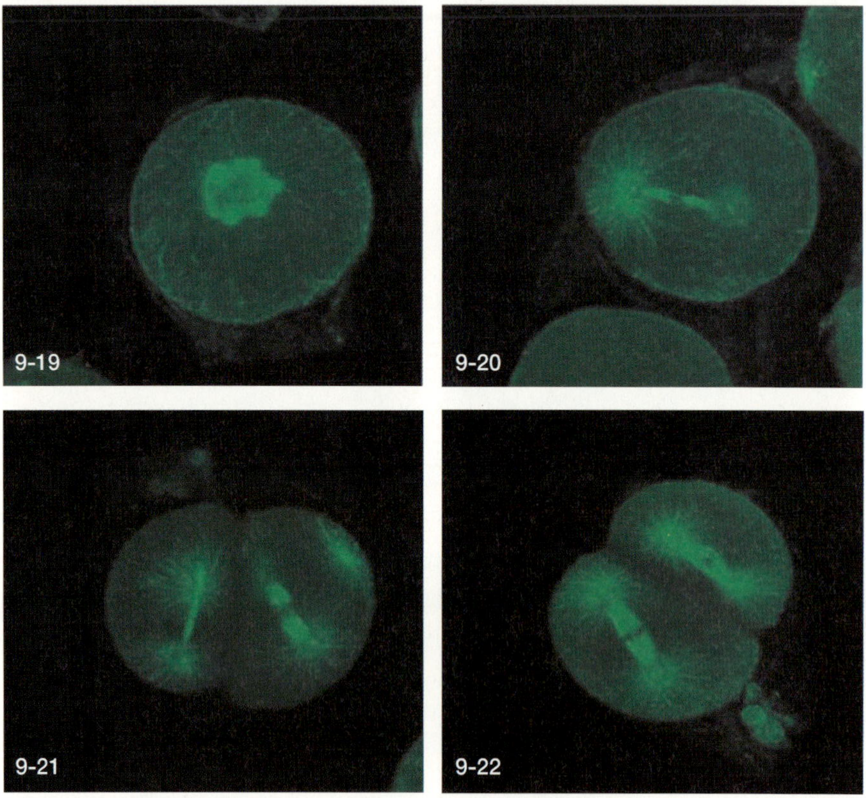

Fig. 9-19 Sperm is added, but the germinal vesicle is not yet break-downed. Anti-α-, -γ-tubulin active cytoskeletal system connects germinal vesicle to oolemma in the oocyte. × 400.
Fig. 9-20 In proceeding of the first meridional cleavage, the mitotic apparatus is arranged perpendicular to animal-vegetal axis. Polarity of the bipolar spindle fiber poles is differed from the other pole. × 400.
Fig. 9- 21, 9-22 In proceeding of the second meridional cleavage, the spindle fibers are first oriented parallel each other, in advance of time, tilted obliquely opposite direction to one another. × 400.

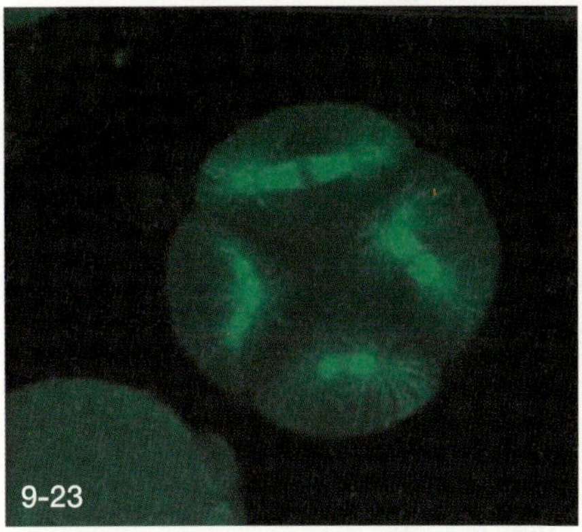

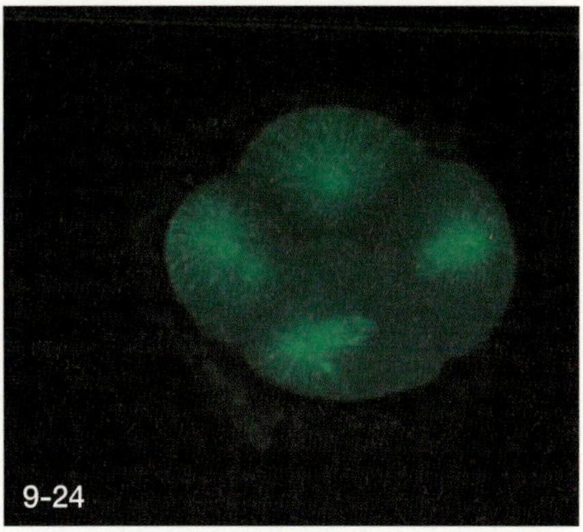

Fig. 9-23, 9-24 The third cleavage is equatorial and the mitotic spindles of opposite blastomeres are aligned parallel with each other, however, as seen in Fig. 9-24, the mitotic spindles of each blastomere rotated clockwise inducing in consequence equatorial cleavage. × 400.

cal to animal-vegetal axis (Fig. 9-23). As the third cleavage is initiated, the MtAs tilted obliquely at a dextral angle (Fig. 9-24). The clockwise tilting of the MtAs determined the ultimate oblique angle of the blastomeres. Consequently, the upper four blastomeres are displaced at a 45 degree angle to the right side to the lower blastomere (Fig. 9-11).

Symmetric and holoblastic cleavage continues until the completion of the fifth cleavage (Fig. 9-12, 9-13). Differences in blastomere size are observed at the completion of the sixth cleavage. The size-differences may come from asynchronous cleavage between the blastomeres of the animal pole site and the vegetal pole site (Fig. 9-14). The marginal blastomeres are larger than the blastomeres at the central portion of the embryos (Fig. 9-14, 9-15). The embryos developed into blastulas, hatched into trocophore larvae equipped with circularly-arranged cilia (Fig. 9-16, 9-17), and finally developed into pigmented worms (Fig. 9-18).

9-3 Cleavage pattern of *Urechis* embryo

Cell cleavage is important as a method for the transformation of unicellular organisms into multicellular organisms. During cleavage, similar cells are grouped into tissues. Various groups of cells or tissues then form organs, and cooperative organs form an individual organism.

Among animal species, cleavage patterns and grouping of the similar cells are commonly divided into five categories: radial, bilateral, rotational, superficial, and spiral cleavages according to the formation of cleavage plans and geometric features of the dividing cells or blastomere. The cleavage plan is commonly determined by Hertwig's rules: the amount of yolk and the orientation of the mitotic apparatus within the non-yolky cytoplasm. The mitotic apparatus migrates to the non-yolky cytoplasm and the cell is cleaved perpendicular to the spindle fibers (Harvey, 1935; Rappaport & Epstein, 1965; Rappaport, 1976; Kuriyama *et al*., 1986; Kaltschmidt & Brand, 2002) or the connective line of the asters (Harvey, 1935; Hiramoto, 1971).

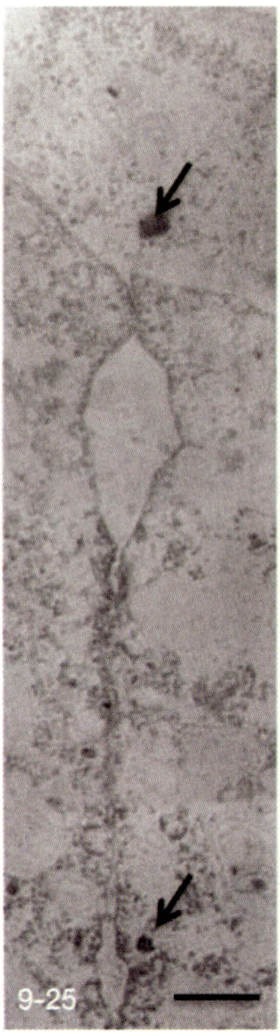

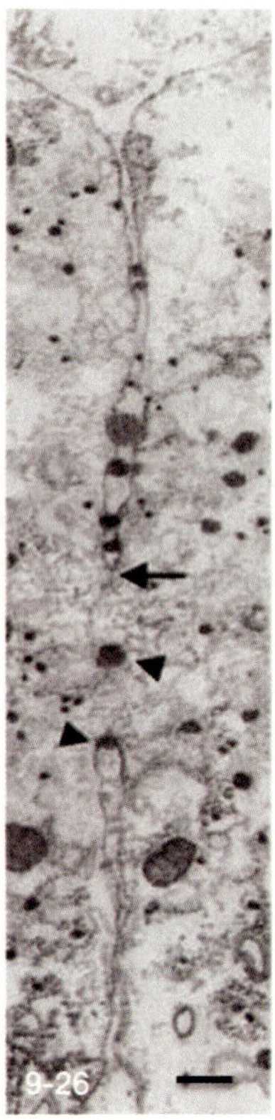

Fig. 9-25 Mosaic picture in reference of many EM micrographs reveals cleavage furrow and contractile ring appeared in proceeding of the first cleavage. Dense inclusion (arrow), aggregate of fine particle are frequently observed at near of the cleavage furrow. × 5,500.

Fig. 9-26 The mosaic picture is redrawn to get a generalized concepts on the whole cleavage furrow and contractile ring. Dense inclusion of various sizes show a tendency to be appeared along the prospective direction of the cleavage furrow. At a late phase of the cleavage, the dense aggregates are dispersed and merged into the plasma membrane of the blastomeres. Arrow head: dense inclusion, Arrow: dispersed and merged particles in blastomere membrane. × 5,500.

In *U. unicinctus,* the oocyte is isolecithal, the yolk materials are distributed evenly in the ooplasm, and the germinal vesicle is located in a slight eccentric ooplasm to the animal pole. Based on previous reports and Hertiwig's rules, isolecithal oocytes are predicted to undergo radial, bilateral, rotational, and spiral cleavage (Browder, 1991).

U. caupo oocytes have been briefly reported to undergo radial cleavage (*U. caupo*: Newby, 1986; Gould-Somero, 1975). As above, except oblique tilting of MtAs, other features of the *Urechis* cleavage are similar to radial cleavage. The features of the radial cleavage are asymmetry of the heart-shaped blastomere of the first cleavage and equal and holoblastic cleavage at subsequent cleavages, as reported in Echinoderm (Okazaki, 1975). The asymmetry of the animal-vegetal hemispheres in the heart-shaped blastomere is disappeared in progress of the cleavage and symmetry is restored completely at the end of the first cleavage. The restoration of symmetry is accomplished by new cleavage furrow from the vegetal pole. Beginning from the vegetal pole, the cleavage furrow proceeded to animal pole and finally encountered the cleavage furrow derived from the animal pole.

Although *Urechis* cleavage is similar to the radial cleavage during the initial steps of cleavage, orientation of the MtAs and arrangement of the blastomeres are decisively different from the radial cleavage. In *U. unicinctus,* the upper and lower blastomeres are organized into a spiral pattern. The oblique tilt of MtAs is alternative, dextral and sinistral tilting is observed according to corresponding mitosis.

The first three cleavages are series of meridional-meridional-equatorial cleavages that produce eight blastomeres of equal size. The blastomeres are in an oblique spiral arrangement. At the completion of the first cleavage, one upper blastomere is displaced obliquely to the right of the lower blatomeres. At the four cell stage, the two upper blastomeres of the opposite sites are displaced obliquely to the left side of the lower blastomeres with respect to the animal-vegetal axis. At the completion of the third cleavage, the four blastomeres at the animal pole are displaced dextrally until they settle into a diagonal position with respect of the four lower blastomeres. Exclusive of the equal size of eight blastomere, the diagonal displacement of the upper four blastomeres are similar to arrangement of the eight blastomeres in the mollusks (Reverberi, 1971; Verdonk & Biggelaar, 1983), annelids (Freeman & Lundelius, 1982; Verdonk & Biggerlaar, 1983), some flatworms (Reververi, 1971), and nemertine worms (Gilbert, 1988).

Oblique tilting of mitotic apparatuses (MtAs) is the primary difference between spiral cleavage and all other forms of cleavage (Harris, *et al.*, 1980; Sluder *et al.*, 1989; Hamaguchi, 2001; Verdonk & Biggerlaar, 1983). Anti-tubulin immunofluorescence staining revealed the oblique tilting of the MtAs during every mitotic cycle of *Urechis* cleavage. During the first meridional cleavage of the *Urechis* blastomere, the MtA is tilted dextrally with respect to the animal-vegetal axis, as is predicted in the *U. caupo* zygote (Newby, 1986; Gould-Somero & Hollander, 1975). At the end of the first cleavage, the blastomeres are slightly displaced as described above. In proceeding of the second cleavage, the mitotic apparatuses of the two blastomeres are tilted sinisterly opposite to one another. Tilting of MtAs in the third cleavage continue until subsequent equatorial cleavage formed the eight blastomeres of equal size. The third MtAs of the blastomeres facing one another initially appeared parallel to each other. However, the MtAs of four blastomeres inclined dextrally during the third cleavage until they formed a complete oblique angle. Consequently, the upper four blastomeres are arranged diagonally to the lower four blastomeres. The inclination angles of the third MtAs exceeded the tilted angles of the previous cleavages. The upper blastomeres of the first and second cleavage slightly overlapped the lower blastomeres, but after completion of the third cleavage, the alignment of the upper four blsatomeres exceeded the overlapping angle of the two or four blastomeres and approached the complete oblique angle.

In many respects, spiral cleavage and arrangement of the *U. unicinctus* blastomeres have unique features compared to spiral cleavage in other organisms. First, the timing and stage of the diagonal blastomere displacement are similar to that of mollusks (snail: Verdonk & Biggelaar, 1983), but different from annelids in which the diagonal arrangements appear early at the two-cell stage (*Arenicola,* Reverberi, 1971). Second, unique characteristic of the *Spiralian* cleavage is known as unequal cleavage producing unequal sized blastomeres, which are appeared early at two-cell stage, or at least until the four-cell stage in most *Spiralian* (Schroeder, 1973; Del-Pino, 1989). In contrast of the unequal cleavage, symmetric and holoblastic cleavage continued at completion of the fifth cleavage in *U. unicinctus*. The size differences between blastomeres are observed at end of sixth cleavage. However, there is no macro-, meso-, and micromere tiers to be observed. The size differences may come from the asynchronous cleavages between blastomeres at animal pole and vegetal pole. The cleavage of blastomeres at vegetal pole is caused to be slowed down comparing with the blastomeres at animal pole. In other words, the MtAs are tilted during every mitosis cycle as like other *Spiralian,* but the macro-, meso-, and micromere tiers are not exactly formed in *Urechis* embryo. On the other hand, retraction of the polar bodies into the presumptive macromere (*Ilyanassa:* Conrad, 1973; Dohmen, 1983) is unlikely to occur in *U. unicintus*, thereby retaining equal cleavage until end of the fifth cleavage.

In addition, *U. unicinctus* oocytes are fertilized at the germinal vesicle stage prior to enter meiotic cycle. *Urechis* oocytes enter meiotic cycle after fertilization. Staining with anti-tubulin-FITC conjugate revealed that the meiotic apparatuses (Maps) of *U. unicinctus* rotates and shifts during the first and second meiotic cycles (Kwon, *et al.*, 2003, Shin *et al.*, 2005a, Shin *et al.*, 2005b). The Maps appear in the middle of the zygote, migrate to the prospective site of oolemma, from which polar bodies extrude from the zygote. The Maps are located parallel to the tangential line of the corresponding oolemma, but rotates vertically to the tangential line of the oolemma when the polar bodies extrude from the zygote. The Maps are rotated and shifted during migration from the appearing site to functional site. The initial orientation of the meiotic apparatus is different from its orientation in the action site.

During initial cleavage, contractile ring at the leading edge of the cleavage furrow is observed in *Urechis* embryo. Under EM, fine structures and thickened region at the frontal tip of the first cleavage furrow is observed in *U. unicinctus* (Shin *et al.,* 2004). The fine structure of *Urechis* contractile ring is similar to contractile ring known in echinoderm (Szollosi, 1970). The contractile ring pinches the zygote into two parts through a mechanism similar to muscle contraction (Schroeder, 1973). The contractile ring is positive in antibody staining against actin and myosin, and the positive reaction is proved by treatment of cytochalasin B (Asnes, 1979; Schroeder, 1987). In *Urechis,* dense body 7~10 nm in size is incorporated into the leading edge of the cleavage furrow, and microvillous structures are observed in the deep apical region of the cleavage furrow. The presence of the microvill implies that the region is derived from the oocyte surface, even though the region deeply penetrates into the ooplasm. Most of dense bodies initially appear to be 7~10 nm in size. A number of dense bodies are found in a row of cleavage furrows. It is likely that the dense bodies of the ooplasm are initially arranged in a row along the prospective direction of the cleavage furrows. Once the cleavage furrows pass through the location of the dense bodies, the dense bodies are dispersed and emerge on the plasma membrane of each blastomere. Dense bodies are also observed at the center of zygotes where both cleavage furrows meet and fuse together to finish the cleaving process.

Summary

Oocyte of *U. unicinctus* is isolecithal, and the germinal vesicle is located at a slight eccentric position to the animal pole, it matters little to say that the germinal vesicle is located in the middle of oocyte. Isolecithal oocytes are predicted to undergo radial, bilateral, rotational, and spiral cleavage. It reveals cleavage pattern of *U unicinctus* oocyte is spiral cleavage.

U. unicinctus oocyte shows a number features of radial cleavage. The features are equal holoblastic cleavage and appearance of heart-shaped blastomere during the first cleavage. Even though the features of radial cleavage appeared, anti-tubulin-FITC immunofluorescence incorporations reveal the oblique tilting of the mitotic apparatuses during every mitotic cycle. During the first meridional cleavage, the mitotic apparatus is tilted dextrally with respect to animal-vegetal pole. In progress of the second meridional cleavage, the mitotic apparatus tilts sinisterly. The most distinctive tilting of the mitotic apparatuses is observed in progress of third cleavage producing eight blastomeres. The mitotic apparatuses of the third cleavage are appeared at first parallel to animal-vegetal axis, but tilts dextrally until the tilting reaches complete diagonal angle, so that the subsequent equatorial cleavage formed eight blastomeres of equal size. As the results, upper four blastomeres are arranged diagonally to the lower four blastomeres. The inclination angles of the third mitotic apparatus exceeded the tilted angles of the previous cleavages.

Sequential cleavages producing equal and holoblastic blastomere at least until the fifth cleavage are differences from typical spiral cleavage. In the most *Spiralian,* unequal cleavage and differences in blastomere size appear early at two-cell stage, or at least until the four-cell stage. In *U. unicicntus,* symmetric and holoblastic cleavage continued at completion of the fifth cleavage.

Under EM, fine structures and thickened region at the frontal tip of the first cleavage furrow is suggested as presence of contratile ring. Dense body 7~10 nm in size is incorporated into the leading edge of the cleavage furrow, and microvillous structures are observed in the deep apical region of the cleavage furrow. The dens bodies are concentrated at appearance, dispersed and emerged mainly on the plasma membrane of each blastomere.

REFERENCES

Asnes CF, TE Schroeder: Cell cleavage: Ultrastructural evidence against equatorial stimulation by aster microtubules. Exp Cell Res, 122: 327-338, 1979.

Bestor TH, G Schatten: Anti-tubulin immunofluorescence microscopy of microtubules present during the pronuclear movements of sea urchin fertilization. Dev Biol, 88: 80-91, 1981.

Browder,LW, CA Carol, WR Jeffrey: Cleavage: Becoming multicellular. In " Developmental biology", 3rd ed. Saunder Colleg Pub. pp 163-195, 1991.

Conrad G: Control of polar lobe formation in fertilized eggs of *Ilyanassa obsolete* Stimpson. Amer J Zool, 13: 961-980,1973.

Davis FC: Changes in RNA polyadenylation and protein synthesis at fertilization in *Urechis caupo*. J Cell Biol, 103(5) part 85A, 1986

Del-Pino EM: Modifications of oogenesis and development in marsupial frogs. Development, 107: 169-187,1989.

Dohmen MR: The polar lobe in eggs of mollusks and annelids: Structure, composition, and function. In WR Jeffery and RA Raff ed, *Time, space and pattern in embryonic development.* Alan R. Liss, New York, pp197-220, 1983.

Engstrom WS: Removal of the fertilization membrane of fertilized eggs of *Urechis caupo* and development of membraneless embryos. Biol Bull, 140(3): 369-375, 1971.

Freeman G, JW Lundelius: The developmental genetics of dextrality and sinistrality in gastropod *Lymnaea peregra*. Wilhelm Roux's Archives, 191: 69-83, 1982.

Gilbert SF: Developmental Biology, 2nd ed. Sinauer Associates, Sunderland, MA, 1988.

Gould-Somero M: Echiura. In: reproduction of Marine Invertebrates. Vol. III, Academic Press, New York, pp 277-311,1975.

Gould-Somero M, L Hollander: Fine structural investigation of the insemination response in

Urechis caupo. Dev Biol, 46: 358-369, 1975.

Hamaguchi Y: Displacement of the mitotic apparatus which induces ectopic polar body formation or parthenogenetic cleavage in the star fish oocytes. Dev Biol, 364: 239-375, 2001.

Harris P, M Osborn, K Weber: Distribution of tubulin containing structures in eggs of the sea urchin Strogylocentrotus purpuratus. J Cell Biol, 84: 668-679, 1980.

Harvey EB: The mitotic figure and the cleavage plane in the egg of *Parechinus microtuberbulatus*, as influenced by centrifugal force. Biol Bull, 69: 287-297, 1935.

He X, M Dembo: On the mechanics of the first cleavages division of the sea urchin egg. Exp Cell Res 233: 252-273, 1997.

Halanych KM: The phylogenetic position of the pterobranch hemichordates based on 18S rDNA sequence data. Mol Phylogenet Evol ,4(1): 72-76, 1995.

Hiramoto Y: Analysis of cleavage stimulus by mean of micromanipulation in sea urchin eggs. Exp Cell Res, 68: 291-298, 1971.

Kaltschmidt JA, AH Brand: Asymmetry cell division: microtubule dynamics and spindle asymmetry. J Cell Sci 115(11): 2257-2264, 2002.

Kitajima A, Y Hamaguchi: Determination of the first cleavage plane: the relationships between the orientation of the mitotic apparatus for th e first cleavage and the position of meiotic division-related structures in starfish eggs. Dev Biol 280: 48-58, 2005.

Kuriyama R, GG Borisy, Y Masui: Microtubule cycles in oocyte of the surf clam, Spisula solidissima: An immunofluorescence study. Dev Biol, 114: 151-160, 1986.

Kwon Hj, JW Jeong, WJ Kim, KS Shin: Immunofluorescence and electron microscopic study on the artificial insemination and rotation-shift behaviors of the bipolar spindle fibers in *Urechis unicinctus* egg. Korean J Electron Microscopy, 33(2): 105-116, 2003.

McHugh D: Molecular evidence that echiurans and pogonophorans are derived annelid. Proc Natl Acad Sci USA, 94: 8006-8009, 1997.

Muneoka Y, M Kamura: Actions of noradrenalin and some other biogenic amines on the body wall muscles of an echiuorid *Urechis unicinctus*. Comp Biochem Physiol. Ciomp Pharmacol, 72(2): 281-287, 1982

Newby WW: The embryology of the echiuroid worm *Urechis caupo*. Mem Amer Phil Soc, 16: 1-19, 1940.

Newby WW: The embryology of the echiuroid worm *Urechis caupo*. Mem Amer Phil Soc, 16:1-219, In "Reproduction of marine invertebrates." Vol III, 118: 148-154, 1986.

Nielsn C: Animal evolution: interrelationships of the living phyla. Oxford Univ, Press Oxford, pp 1-467, 1996.

Okazaki K: Normal development to metamorphosis. In "The sea urchin embryo: Biochemistry and morphogenesis." G Czihiak ed, Springer Verl, pp 177-232, 1975.

Rappaport R, RP Epstein: Duration of stimulus and latent period proceeding furrow formation in sand dollar eggs. J Exp Zool, 158: 373-382,1965.

Rappaport R: Furrowing in altered cell surfaces. J Exp Zool, 195: 271-277, 1976.

Reverberi G: Annelids. In G Reverberi ed, Experimental embryology of marine and freshwater invertebrates. Elsevier, North Holland Biomedical press, Amsterdam, pp 126-163, 1971.

Sato H: Studies on the echiuroidea, sipuncluoidea and priapuloidea of Japan. Sci Rep Tohoku Imp Univ Ser IV vol XIV, Pl XXIII, pp339-460, 1939.

Schroeder TE: Cell constriction: Contractile role of microfilaments in division and development. Amer Zool, 13: 949-960, 1973.

Schroeder TE: Fourth cleavage of sea urchin blastomeres: Microtubule patterns and myosin localization in equal and unequal cell divisions. Dev Biol, 124: 9-22, 1987.

Shin KS, DH Lee, TY Ko: Cleavage pattern of Urechis unicinctus egg in vitoro fertilization. Korean J Electron Microscopy, 34(1): 71-81, 2004.

Shin KS, HJ Kwon, WJ Kim: Attribution to cortical granules to formation of fertilization envelope and polyspermy block in *Urechis unicinctus*. Integrative Biosciences, 9(2): 57-64, 2005b.

Shin KS, Hj Kim, HJ Kwon, WJ Kim: Cytological localization of nuclear actin of sperm and spermatids in *Urechis unicinctus*. Integrative Biosciences, 9(2): 65-73, 2005a.

Sluder G, FJ Miller, K Lewis, ED Davidson, CL Rider: Centrosome inheritance in star fish zygote: Selective loss of the maternal controsome after fertilization. Dev Biol, 131: 567-579, 1989.

Szollosi D: Cortical cytoplasmic filaments of cleaving eggs: A structural element corresponding to the contractile ring. J Cell Biol, 44: 192-209, 1970.

Tsutomu K, H Hiroshi, O Syozo: The nucleotide sequences of 5s RNAs from two annelid species *Perinereis brevicirris* and *Sabellastarte japonica* and an echiuran species *Urechis unicinctus*. Necleic Acid Res, 11(10): 3347-3350, 1983.

Verdonk NH, JAM van den Biggerlaar: Early development and formation of germ layers. In "The mollusk." Vol 3, Development, K Wilbur ed, Academic Press, New York, pp 91-122, 1983.

Chapter 10

Hox genes of the echiuroid *Urechis unicinctus*

Chapter 10
Hox genes of the echiuroid *Urechis unicinctus*

10-1 Introduction

Genes containing a homeobox form a large family of genes that encode transcription factors. These genes, particularly *Hox* genes, play a pivotal role in the determining the axial properties of many animal species (Scott *et al*., 1989; Gehring *et al*., 1990). Reconstructing the evolution of *Hox* gene organization is expected to provide valuable information about animal phylogeny. Amino acid sequences within the homeodomain proteins are highly conserved through evolution and are distributed over a wide phylogenetic spectrum (Balavoine and Telford 1995; Gehring 1993).

Molecular phylogenetic analyses have resulted in significant advances in constructing the metazoan evolutionary tree. Since researchers have begun to apply *Hox* cluster and 18S rRNA analysis to metazoan phylogeny studies, remarkable progress has been made towards understanding evolutionary relationships (Halanych, 1995; Nielsn, 1996; McHugh, 1997; Balavoine *et al.,* 2002). In particular, the use of *Hox* gene analysis has confirmed some traditional hypothesis based on morphology or embryology, such as the monophyly of bilateralians (de Rosa *et al.,* 1999; Balavoine *et al.,* 2002).

Annelida, which includes the polychaetes and the clitellates, has long held the taxonomic rank of phylum. Unsegmented, mud-dwelling echiurans and gutless, deep sea pogonophoran tube worms share several embryological and morphological features with annelids, but each group is also considered to be a separate metazoan phylum based on the unique characteristics exhibited by each group (Rouse & Fauchald, 1997).

Echiurans are plump marine worms that possess a flexible proboscis at the end of the mouth. They were classically grouped in a separate phylum due to their unsegmented, coelomate bodies in addition to their unique anal sacs, proboscis, and muscle layer (Newby, 1940). However, recent molecular data comparing the sequences of the elongation factor-1 genes indicate that this taxon should be groups with annelids (McHugh, 1997). Three gene fragments of the echiuroid *Hox* / non-*Hox* genes (*Hox 3, EMS*, and *Gbx2*) have been isolated from *Urechis caupo* (Lee *et al*., 2003). In the present study, we determined that the common echiuroid species, *Urechis unicinctus,* possess three anterior, five central, and one posterior group of *Hox* genes.

10-2 Homeodomains in *U. unicinctus*

By performing PCR with degenerate primers, we were able to isolate three *Hox* fragments from *U. unicinctus* genomic DNA: *Uun-lab01* and *-lab02* (amplified with the P03/P02 primer pair), *Uun-Hox3* (P01 and P02 primers), and *Uun-Post2* (P06 and P02 primers). To obtain more sequence information, reverse transcription-PCR was performed using specific degenerate primers designed based on hexapeptide motif sequences (Table 1). Using this set of primers, cDNA was amplified from larvae of various stages. Because the 5'-flanking sequence of a homeodomain is usually an

Table 1. PCR primers use to amplify echiuroid *Hox* genes

Target-specific primer sequence (5' to 3') corresponding amino acid sequence, names of the genes, and position in the homeodomain
P01 Hox universal Forward-YTNGARYTNGRARAARGARTTY LELEKEF/14-20
P02 Hox universal* Reverse-NCKNCKRTTYTGRAACCADATYTT KIWFQNRR/46-53
P03 labial Forward-ACCAACTTYACHAAYAARCAR TNFTNKQ/6-12
P04 Lox6, Scr, Anp Forward-ATHTAYCCNTGGATG IYPWM/N-terminal to homeodomain
P05 Lox2, Lox4 Forward-CCNCARATHTAYCCNTGGATGPQIYPWM/N-terminal to homeodomain
P06 Post Forward-CARMGWAARAARMGWAARCCQRKKRKP/1-7

*The P02 primer served as a reverse primer for all *Hox* genes informative clone, we were able to amplify nine unique *Hox* gene fragments in addition to non-*Hox* cluster genes such as *Xlox* and *Gbx* (data not shown).

Analysis of the allelic variants as based on pairwise comparisons of the number of nucleotide substitutions within corresponding region of paralogous *Hox* genes (Irvine *et al*., 1997; Mito and Endo, 2000). All pairwise comparisons of the *U. unicinctus* gene fragments involved 17 or more nucleotide substitutions and sequence similarities less than 77%. Each unique *Hox* fragment was named with the prefix *Uun* for *U. unicinctus* followed by a specific identifier which reflected *Hox* gene orthology and, if necessary, a serial number (*e.g.*, *Uun- lab01*).

Our PCR survey using degenerate primers specific for each PG allowed us to more clearly identify *Hox* gene orthology, particularly in the central group genes. Nine PGs were consisted of three anterior group genes (*labial*/PG1, one PG3) and five central group genes (*Post2*) and one posterior group of *Hox* genes. Fig. 10-1 and 10-2 show the alignment of *U. unicinctus Hox* fragments and known *Hox* gene sequences of other lophotrochozoans and *Drosophil*a. We were able to amplify three distinct anterior group genes among which two fragments were assigned to PG1 and one to PG3.

The relationships of Uun-lab01, Uun-lab02, and Uun-Hox3 with anterior *Hox* genes of other lophotrochozoans were clearly illustrated by the characteristic positions of the amino acid substitutions in the anterior *Hox* genes as shown in Fig. 10-1. The putative homeodomain sequences of *U. unicinctus* orthologs, Uun-lab01 and -lab02, contained PG1-specific A29, N41, and T43 residues (Canapa *et al*., 2005) and appear to be very similar to the PG1 homeodomains of oligochaeta (Pex-lab01, Pex-lab02, and Pex-lab03), polychaeta (Nvi-lab), cephalopods (Esc-lab), brachiopods (Lan-lab), and leeches (Htr-Lox7). In particular, the Uun-lab01 fragment encoded a peptide with 100% amino acid sequence identity with the labial orthologs from oligochaeta Pex-lab01 (Fig. 10-1). Although we were able to identify two types of PG1 genes in *U. unicinctus*, the sequence of Uun-lab01 and Pex-lab01 (Fig. 10-1) showed the highest amino acid and nucleotide similarities. Further investigation is required to verify the origin of these genes. However, it was also noted that multiple labial orthologs

(PG 1) have been reported in other lophotrochozoans, *Stylaria lacustris*, and *P. excavatus* (Snow and Buss 1994; Cho *et al*., 2003).

Uun-Hox3 was assigned to PG3. This fragment showed a significant similarity with PG3 members of representative lophotrochozoans species including the oligochaetae *P. excavatus* (*Pex-Hox3*), cephalopod *E. scolopes* (Esc-Hox3), polychaetae *N. virens* (*Nvi-Hox3*), bryozoans (*Btu-Hox3*), and brachiopod *L. anatina* (Lan-Hox3). In particular, the conceptually translated *Uun-Hox3* was 100% identical to that of *Lan-Hox3* isolated from the brachiopod *L. anatina* (Fig. 10-1).

The *Uun- Hox3* fragment (Fig. 10-1) that contained 25 amino acid residues within the homeodomein (21-24) contained one residue, N37, which is specific to Hox3/PG3 (Canapa *et al*., 2005).

Five distinct fragments were recovered and each was assigned to one of the five central PGs based on the conserved residues within the homeodomain. Lophotrochozoans are characterized by the presence of *Lox5*, *Lox2*, and *Lox4* in the central group (Balavoine *et al*., 2002). The present study provides evidence of the presence of *Lox2* and *Lox4* in *U. unicinctus*, but *Lox5* was not amplified. Additionally, there appear to be at least two *Dfd* orthologs in *Lox6* and *Lox18* of leeches (Kourakis and Martindale, 2001); however, only one *Dfd* ortholog was recovered in echiuroids. *Uun-Lox6* showed a 96% amino acid identity with *Pex-Lox6* from *P. excavatus*.

Uun-Scr is a PG5 ortholog, identified by the Sex combs rescued (Scr) gene. Uun-Scr displayed the highest similarity (Fig. 10-1) to Nvi-Scr (96%) and Lan-Scr (96%). Uun-Scr (Fig. 10-1) was found to contain highly conserved two residues at specific positions (H11 and N39) that provide a signature for Scr/PG 5 homologs (Giusti *et al*., 2000). Uun-Antp (Fig. 10-1) was significantly similar to Antenneapedia orthologs from other lophotrochozoans such as oligochaete (Pex-Antp), polychaete (Nvi-HB1), cephalopods (Esc-Antp), and brachiopods (Lan-Antp).

The *Lox2* and *Lox4* genes of lophotrochozoans are thought to be independent duplications of the ecdysozoan *Ubx* and *AbdA* genes (Balavoine *et al*., 2002). *Uun-Lox2* was identified by *Lox2*-characteristic residues at K21 and S35 (Fig. 10-1) and is most similar to *Lox2* from the leech *H. medicinalis* (Balavoine *et al*., 2002). The Uun-Lox4 amino acid sequence was identical to that of *Lox4* from the leech *H. medicinalis* (Fig. 10-1) and contained three conserved residues, S9, Q21, and K29 (Balavoine *et al.* 2002).

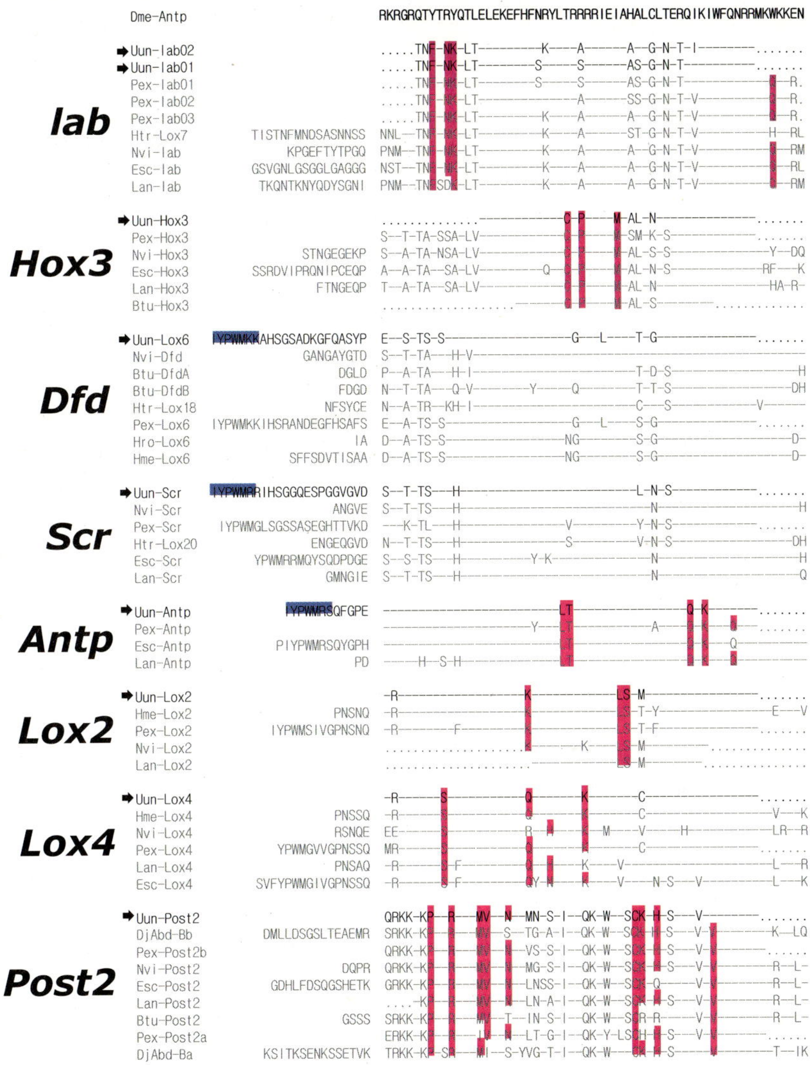

Fig. 10-1 Alignment of the amino acid sequences of the homeodeomains. The sequences of *U. unicinctus* (Uun-) indicated by arrows are aligned in comparison to those from other lophotrochozoans and *Drosophila* homeodomains. Dashes indicate identity to the top reference sequence (*D. melanogaster* Antp). Amino acids that are characteristic of each orhtology-specific residues are shaded.

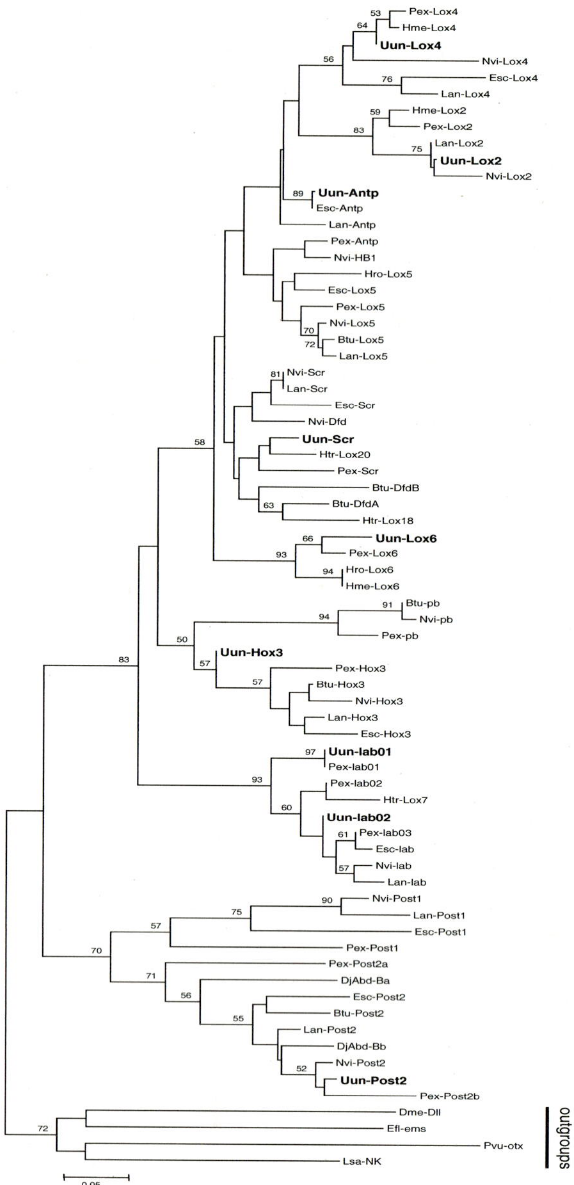

Fig. 10-2 Phylogenic relationship among homeodomain sequences based on neighbor-joining analysis of the deduced amino acid sequences. The number at the nodes are scores from 1,000 bootstrap resamplings of the data. Unlabeled nodes are supported by fewer than 50% of bootstrap value. The dataset is as described in "Materials and methods". *Urechis Hox* genes are bolded.

Lophotrochozoans are also characterized by the presence of two posterior genes: *Post1* and *Post2* (de Rosa *et. al.,* 1999). We were able to amplify one distinct fragment determined to be from *Post2* according to the conserved residues within the homeodomain (Fig. 10-1). Comparative analysis (Fig. 10-1, 10-2) revealed that Uun-Post2 had a very high amino acid sequence similarity with the *Post2* genes from other lophotrochozoans such as the oligochaetes (Pex-post2), polychaetes (Nvi-post2), cephalopods (Esc-post2), bryozoans (Btu-post2), and brachiopods (Lan-post2). In addition, Uun- Post2 (Fig. 10-1) appeared to share several conserved amino acids within the homeodomain specific to the lophotrochozoan *Post2* gene located at M14, V15, I26, S35, C36, and K37 (Passamaneck and Halanych 2004).

Although we could not identify orthologs of one of the anterior group, central group and posterior group genes, echiuroids appeared to have lophotrochozoan-specific *Hox* PGs such as *Lox2*, *Lox4*, and *Post2*.

More sequence information and studies are required to determine the genomic architecture of the echiuroid *Hox* genes.

Summary

The *Hox* genes of an echiuroid species, *Urechis unicinctus*, were studied using polymerase chain reaction with homeobox-specific degenerate primers. We identified nine gene fragments containing distinct homeodomains. The nine fragments were classified according to *Hox* genes (three were from the anterior group, five were from the central group, and one was from the posterior group.

REFERERNCES

Balavoine G, MJ Telford: Identification of planarian homeobox sequences indicates the antiquity of most *Hox*/homeotic gene subclasses. Proc Natl Acad Sci USA, 92: 7227-7231, 1995.

Balavoine G, R de Rosa, A Adoutte: *Hox* clusters and bilaterian phylogeny. Mol Phylogenet Evol, 24: 366-373, 2002.

Canapa A, MA Biscotti, E Olmo, M Barucca: Isolation of *Hox* and *ParaHox* genes in the bivalve *Pecten maximus*. Gene, 348: 83- 88, 2005.

Cho SJ, PY Cho, MS Lee, SY Hur, JA Lee, SJ Kim, KS Koh, YE Na, JK Choo, CB Kim, SC Park:

Hox genes from the earthworm *Perionyx excavatus*. Dev Genes Evol, 213: 207-210, 2003.

de Rosa R, JK Grenier, T Andreeva, CE Cook, A Adoutte, M Akam, SB Carroll: *Hox* genes in brachiopods and priapulids and protostome evolotion. Nature, 399: 772-776, 1999.

Gehring WJ: Exploring the homeobox. Gene, 135: 215-221, 1993.

Gehring WJ, M Muller, M Affolter, A Percival-Smith, M Billeter, YQ Qian, G Otting, K Wuthrich: The structure of the homeodomain and its functional implications. Trends Genet, 6(10): 323-329, 1990.

Giusti AF, VF HinmanF, SM Degnan, BM Degnan, DE Morse: Expression of a *Scr/H-ox* gene in the larval central nervous system of the gastropod *Haliotis.* a nonsegmented spiralian lophotrochozoan. Evol Dev, 2(5): 294-302, 2000.

Halanych KM: The phylogenetic position of the pterobranch hemichordates based on 18S rDNA sequence data. Mol Phylogenet Evol ,4(1): 72-76, 1995.

Irvine SQ, SA Warinner, JD Hunter, MQ Martindale: A survey of homeobox genes in *Chaetopterus variopedatus* and analysis of polychaete homeodomains. Mol Phylogenet Evol, 7: 331-345, 1997.

Kourakis MJ, MQ Martindale: *Hox* gene duplication and deployment in the annelid leech *Helobdella.* Evol Dev, 3: 145-153, 2001.

Lee SE, RD Gates, DK Jacobs: Gene fishing: The use of simple protocol to isolate multiple homeodomain classes from diverse invertebrate taxa. J Mol Evol 56: 509-516, 2003.

McHugh D: Molecular evidence that echiurans and pogonophorans are derived annelida. Proc Natl Acad Sci USA, 94: 8006-8009, 1997.

Mito T, K Endo: PCR survey of *Hox* genes in the crinoid and ophiuriod: evidence for anterior conservation and posterior expansion in the echinoderm *Hox* gene cluster. Mol Phylogenet Evol, 14: 375-388, 2000.

Newby WW: The embryology of the echiuroid worm *Urechis caupo*. Men Am Philos Soc, 16: 1-19, 1940.

Nielsn C: Animal evolution: interrelationships of the living phyla. Oxford Univ, Press Oxford, pp 1-467, 1996.

Passamaneck YJ, KM Halanych: Evidence form *Hox* genes that bryozoans are lophotrochozoans. Evol Dev, 6: 275-281, 2004.

Rouse GW, K Fauchald: Cladistics and polychaetes. Zool Sci, 26: 139-204, 1997.

Scott MP, JW Tamkun, GW Hatzell Ⅲ: The structure and function of the homeodomain. Biochim Biophys Acta 989: 25-48, 1989.

Snow P, LW Buss: HOM/Hox-type homeoboxes from *Stylaria lacustris* (Annelida:Oligochaeta). Mol Phylogenet Evol, 3: 360-364, 1994

Index

Index

D

E

F